CHROMOSOMAL AND GENOMIC STRUCTURES AND FUNCTIONS

VOLUME 1

Chromosomes and Cell Division

Chromosomal and Genomic Structures and Functions, 2-volume set:
Volume 1: Chromosomes and Cell Division
Volume 2: Genetic Material Structure and Functions
ISBN: 978-1-77964-028-4 (hbk)

Chromosomal and Genomic Structures and Functions, Volume 1:
Chromosomes and Cell Division
ISBN: 978-1-77964-024-6 (hbk)
ISBN: 978-1-77964-025-3 (ebk)

Chromosomal and Genomic Structures and Functions, Volume 2:
Genetic Material Structure and Functions
ISBN: 978-1-77964-026-0 (hbk)
ISBN: 978-1-77964-027-7 (ebk)

CHROMOSOMAL AND GENOMIC STRUCTURES AND FUNCTIONS

VOLUME 1

Chromosomes and Cell Division

Tariq Ahmad Bhat, PhD
Jameel M. Al-Khayri, PhD
Editors

APPLE ACADEMIC PRESS

First edition published 2026

Apple Academic Press Inc.
1265 Goldenrod Circle, NE,
Palm Bay, FL 32905 USA

760 Laurentian Drive, Unit 19,
Burlington, ON L7N 0A4, Canada

CRC Press
2385 NW Executive Center Drive,
Suite 320, Boca Raton FL 33431

4 Park Square, Milton Park,
Abingdon, Oxon, OX14 4RN, UK

Library and Archives Canada Cataloguing in Publication

CIP data on file with Canada Library and Archives

Library of Congress Cataloging-in-Publication Data

CIP data on file with US Library of Congress

ISBN: 978-1-77964-024-6 (hbk)
ISBN: 978-1-77964-025-3 (ebk)
DOI: 10.1201/9781779640253

Dedication

This Book is Dedicated

To

Professor Dr. Ainul Haq Khan,
Ex Chairman Department of Botany and Dean Life Sciences, Aligarh Muslim University, Aligarh, India, who introduced me to the fascinating field of Genetics.

About the Editors

Tariq Ahmad Bhat, PhD
Senior Lecturer of Botany, Department of Education, Government of Jammu and Kashmir, India

Tariq Ahmad Bhat, PhD, currently serves as a Senior Lecturer in Botany in the Department of Education, Government of Jammu and Kashmir, India. His research focuses on molecular genetics, plant biotechnology, mutation breeding, and the genetic improvement of legumes and medicinal plants. He earned his MSc and PhD from Aligarh Muslim University, Aligarh, India. Dr. Bhat has published 16 international books, 100 research papers, review articles, and book chapters, and has participated in 50 national and international conferences, training programs, and workshops. His books include *Chromosome Structure and Aberrations; Mutagenesis and Crop Improvement; Mutagenesis, Cytotoxicity, and Crop Improvement;* and *Biotechnologies and Genetics in Plant Mutation Breeding* (Volumes 1, 2, and 3). He has also authored *Genetic Engineering* (Volumes 1 and 2) and *Molecular Genetics* (Volumes 1 and 2). With 20 years of active teaching experience, Dr. Bhat is also the District Coordinator (Anantnag) for a prestigious medicinal plant research project funded by the National Medicinal Plants Board (NMPB), New Delhi, under the Ministry of AYUSH, India. Additionally, he is a founding faculty member of the Hon'ble CM's Super 50 initiative, launched by the Government of Jammu and Kashmir. Dr. Bhat is a life member of several scientific societies and serves as a reviewer for multiple international scientific journals. In recognition of his contributions, the Government of India honored him with the Best Innovative Science Teacher Award (2014) for his outstanding and meritorious service. In 2022, he was awarded the Sir Syed Gold Medal and a Certificate of Merit by the Jammu and Kashmir Alig Society for his exceptional contributions to education and research. Dr Tariq Ahmad Bhat was awarded with Asian Admirable Achievers 2025 for his outstanding performance,meritorious service and selfless efforts.

Jameel M. Al-Khayri, PhD

Professor of Plant Biotechnology, Department of Agricultural Biotechnology, College of Agriculture and Food Sciences, King Faisal University, Saudi Arabia

Jameel M. Al-Khayri, PhD, is a Professor of Plant Biotechnology in the Department of Agricultural Biotechnology, College of Agriculture and Food Sciences, King Faisal University, Saudi Arabia. He earned his BS in Biology from the University of Toledo in 1984, followed by an MS in Agronomy in 1988 and a PhD in Plant Science in 1991 from the University of Arkansas, USA. For the past three decades, Dr. Al-Khayri has dedicated his research to date palm biotechnology. He has published over 60 research articles and reviews in international journals and contributed 30 book chapters. He has also served as an editor for several special issues of international journals on topics including date palm, biotechnology, and sustainable agriculture under abiotic and biotic stress. Additionally, he has edited 15 Springer reference books, including *Date Palm Biotechnology*, *Date Palm Genetic Resources and Utilization* (Volumes 1 and 2), *Date Palm Biotechnology Protocols* (Volumes 1 and 2), and *Advances in Plant Breeding Strategies* (9 volumes). Dr. Al-Khayri is an editorial board member and reviewer for several international journals. He has also played an active role in the organizing and scientific committees of international scientific conferences, contributing over 50 research presentations.

In addition to teaching, advising graduate students, and conducting funded research projects, he has held several administrative positions, including Assistant Director of the Date Palm Research Center, Head of the Department of Plant Biotechnology, and Vice Dean for Development and Quality Assurance. Dr. Al-Khayri is an active member of the International Society for Horticultural Science and the Society for In Vitro Biology. He also serves as the National Correspondent for the International Association of Plant Tissue Culture and Biotechnology. Notably, he was a member of Majlis Ash-Shura (the Saudi Arabian Legislative Council) during its fifth session. Currently, Dr. Al-Khayri leads an active research program on date palm biotechnology, focusing on genetic transformation, secondary metabolites, and in vitro mutagenesis to enhance tolerance to abiotic and biotic stress. His research interests also include the role of biotechnology in food security and the impact of global climate change on agriculture.

Contents

Contributors

Muhammad Ajaib
Department of Botany, Mirpur University of Science and Technology (MUST), Mirpur, Azad Jammu and Kashmir (AJK), Pakistan

Faiza Akhtar
Department of Botany, Government College University, Faisalabad, Pakistan

Muhammad Atif
Department of Botany, Government College University, Faisalabad, Pakistan

Aaradhya Bhatnagar
Indrashil University, Rajpur, Gujarat, India

Umbreen Bibi
Government College University, Faisalabad, Pakistan

Khadija Hafeez
Department of Botany, Government College University, Faisalabad, Pakistan

Tanveer Hussain
Department of Botany, Mirpur University of Science and Technology (MUST), Mirpur, Azad Jammu and Kashmir (AJK), Pakistan

Muhammad Ishtiaq
Department of Botany, Mirpur University of Science and Technology (MUST), Mirpur, Azad Jammu and Kashmir (AJK), Pakistan

Qaisar Mahmood
Department of Biology, College of Sciences, Sakhir, Bahrain; Department of Environmental Sciences, Comsats University, Islamabad, Abbottabad Campus, Pakistan

Mehwish Maqbool
Department of Botany, Mirpur University of Science and Technology (MUST), Mirpur, Azad Jammu and Kashmir (AJK), Pakistan

Mubashir Mazhar
Department of Botany, Mirpur University of Science and Technology (MUST), Mirpur, Azad Jammu and Kashmir (AJK), Pakistan

Farah Naz
Department of Botany, Government College University, Faisalabad, Pakistan

Abida Parveen
Department of Botany, Government College University, Faisalabad, Pakistan

Sachchit Patel
Indian Institute of Advanced Research, Koba Institutional Area, Gandhinagar, Gujarat, India

Shagufta Perveen
Department of Botany, Government College University, Faisalabad, Pakistan

Momina Saeed
Department of Botany, Government College University, Faisalabad, Pakistan

Arpit Shukla
Indian Institute of Advanced Research, Koba Institutional Area, Gandhinagar, Gujarat, India

Fiza Talat
Government College University, Faisalabad, Pakistan

Sumaira Thind
Department of Botany, Government College University, Faisalabad, Pakistan

Muhammad Waqas
Department of Botany, Mirpur University of Science and Technology (MUST), Mirpur, Azad Jammu and Kashmir (AJK), Pakistan

Nadia Yasmeen
Department of Botany, Government College University, Faisalabad, Pakistan

Sara Zafar
Department of Botany, Government College University, Faisalabad, Pakistan

Acknowledgment

The editors extend their appreciation to the Deanship of Scientific Research, Vice Presidency for Graduate Studies and Scientific Research, King Faisal University, Saudi Arabia, for supporting this work through Project Grant No.KFU252516.

Preface

The editors take great pleasure in presenting the book *Chromosomal and Genomic Structures and Functions*, comprising two volumes. Volume 1 is subtitled *Chromosomes and Cell Division*. The primary motivation for writing this book was to consolidate both classical and advanced aspects of chromosomal and genomic structures and functions, with special reference to chromosomes and cell division, in one comprehensive resource for beginners, experts, and those venturing into the fascinating field of chromosome and cell division.

This volume consists of 10 chapters: *Biotechnology and Its Applications, Chromosomal Theory of Inheritance, The Chromosome, Mitosis, Meiosis, Structural Changes in Chromosomes, Gene and Cell Division, Sex Determination, Chromosome and Mutations*, and *Chromosome Movements During Cell Division*. These chapters provide in-depth studies on the relationships and functional mechanisms of chromosomes, genes, and biological processes, benefiting organisms in general and humans in particular. The exploration of these biological concepts has had a profound impact on society, contributing to advancements such as improved inheritance patterns, enhanced disease diagnosis, a better understanding of human gene variation, innovations in drug and pharmaceutical production, more sensitive and specific forensic analysis, and the development of improved agronomic traits that have revolutionized crop science, medical sciences, and advanced diagnostics.

All the chapters in this volume have been contributed by young, brilliant biologists and highly experienced cell biologists from various universities and research institutions across the globe. This book is a valuable asset for stakeholders such as undergraduate and postgraduate students, teachers, and researchers in the fields of botany, zoology, cell biology, genetic engineering, genetics, biotechnology, biochemistry, and agricultural sciences.

The authors would like to express profound gratitude to all the contributors whose research and insights have enriched this book. Special thanks are also due to the publishers, artists, and other staff involved in its production, who were always willing to incorporate modifications even in the final stages of publication—an effort that is greatly appreciated.

—Tariq Ahmad Bhat, PhD

CHAPTER 1

Biotechnology and Its Applications

MUHAMMAD ATIF, SHAGUFTA PERVEEN,* SARA ZAFAR, and ABIDA PARVEEN

Department of Botany, Government College University, Faisalabad, Pakistan

*Corresponding author

ABSTRACT

The development of modern biotechnology revolutionized each and every aspect of human life. The industrial revolution and biotechnological products not only save and facilitate the human way of life but also safeguard the environment. We will discuss the applications of biotechnology in this chapter, including defining technologies such as green biotechnology, white biotechnology, red biotechnology, blue biotechnology, and others. The main economic and social benefits that biotechnology could provide are highlighted. The applications of different types of biotechnologies and their potential in various fields of interest, such as agriculture, medicine, disease management, and chemical compounds of industrial importance, are discussed. Green biotechnology is mainly applied in agricultural processes to produce genetically modified (GM) plants. It is also applied in power generation tools, biofuel, and biopolymer production. Red biotechnology is mainly applied in disease diagnosis, tissue engineering, gene diagnosis, medicine and drug production, vaccines, and stem cell therapy. Blue biotechnology refers to the use of marine organisms and freshwater bodies for industrial applications. It is also used in the food and cosmetic industries and biofuel production. White biotechnology is applied in industrial processes and the production of chemical compounds like vitamins and amino acids. Yellow biotechnology is used for drug and food production from insects and environmental maintenance.

Chromosomal and Genomic Structures and Functions, Volume 1: Chromosomes and Cell Division. Tariq Ahmad Bhat and Jameel M. Al-Khayri (Eds)

DOI: 10.1201/9781779640253-1

1.1 INTRODUCTION

Karl Erkey, a Hungarian engineer, used the term "biotechnology" for the first time in 1919 (Verma et al., 2011). It is linked with living organisms, which are utilized for the formation of new products in favor of mankind. Biotechnology is a combined term of two Greek words (bios and techno). Until that time, biotechnology had been defined and redeveloped many times (Amarakoon, 2017). The European Federation of Biotechnology defined it as "the integration of microbiology, biochemistry, and engineering sciences to attain industrial (technological) applications of microorganisms, cultured tissue cells, and their parts is referred to as biotechnology" (Myshak, 2018).

Biotechnology is considered an interdisciplinary and multidisciplinary field that plays a vital role in a range of processes and products in the life sciences. This technology has been developed for new medicines, better crop and animal production, and the manufacturing of products for daily use. To distinguish the various applications, a color scheme was adopted. Green, red, white, blue, violet, and brown biotechnology will be explained in the later sections of this chapter.

1.2 HISTORICAL BACKGROUND

The centuries years ago, the biotechnology field began and continued to date. There are three main stages that were recognized according to historic events (Verma et al., 2011): (i) ancient biotechnology; (ii) classical biotechnology; and (iii) modern biotechnology. This term also meets with the roots of ancient civilizations from the initiation of the Egyptian and Indus Valley civilizations, about 5000 to 10,000 BC, for the better selection of animals and plants for yield and taste (Goudie, 2018). These plants were produced by vegetative and sexual propagation (Buchholz and Collins, 2013). Microorganisms were used to prepare cheese, butter, beer, wine, and bread (George et al., 2008) and for the treatment of wounds or ailments by using herbs. All that period was described as ancient biotechnology in the history of biotechnology. The traditional biotechnology techniques were used by earlier civilizations in a trial-and-error manner. These activities were only based on typical practices, and discoveries were also based on observations of nature rather than developments.

The discovery of fermentation between the middle of the 19th century and the 1970s is considered to represent the beginning of the next or second phase in biotechnology history, often termed conventional or classical biotechnology (Amarakoon, 2017). At that time, the widespread use of ancient methods such as fermentation and its adaptation for the industrial preparation of enzymes, vinegar, amino acids, and vitamins emerged (Behera et al., 2019). The mission of the institutes formed by Koch, Pasteur, and Lister was to study fermentation and other microbial activities (Smith, 2012). Mendel's groundbreaking research on the basic principles of heredity revolutionized genetics and ushered in the era of carefully regulated plant breeding techniques (Bud, 1993).

The 1950s saw the discovery of the DNA structure, ushering in the contemporary era of biotechnology. The transition of biotechnology was primarily aided by the contributions of two approaches: (i) rDNA technology; and (ii) monoclonal antibody or hybridoma technology. In earlier technology, DNA from different sources was modified to create a brand-new synthetic molecule. By transferring genetic information from one species to another utilizing rDNA technology, crops and animals can be produced with features that are distinct from those attained through conventional breeding methods (Chaudhuri, 2013). On the other hand, hybridoma technology made it possible to produce antibodies against antigens (Zhang, 2012). Both methods are capable of industrial uses very quickly, which sped up the development of biopharmaceuticals as well as diagnostic techniques in the domains of parasitology, virology, and cancer (Ling, 1983) and in the pharmaceutical industry (Walsh, 2005).

1.3 THE COLORED BIOTECHNOLOGY

The challenges of health, food production, and energy have become serious concerns for the whole world. There is a strict need for primary health care for people. Various solutions to respond to these challenges have been established, including advances and developments in biotechnology that could upgrade bioactive molecules discovered in various medicinal plants (Abdin, 2007). The primary four forms of biotechnology have been studied up to this point: industrial biotechnology, also termed white biotechnology; blue biotechnology, which includes the aquatic ecosystem; red biotechnology (industrial biotechnology); and green biotechnology, also known as agricultural and medical biotechnology, respectively. These innovations advance the creation of therapeutic solutions for both individuals and the environment because they are all derived from biological products and systems.

1.3.1 GREEN BIOTECHNOLOGY (AGRICULTURAL)

Green biotechnology aims to counter hunger and is considered the next step in the green revolution. More resistant and fertile plants are produced using this technology, which ensures the application of eco-friendly fertilizers and biopesticides. Among various technologies, tissue culture is the major one. This includes the multiplication and conservation of elite varieties and methods to produce secondary metabolites (bio-farming). Utilizing genetic engineering, this method is also used to select plants and animals with the required features. The reverse breeding and double haploid concept is adapted to produce homozygous plants from heterozygous plants, which comprise desirable traits (Kafarski, 2012). A transgenic technology is adapted to upgrade the vital secondary metabolite production. These hairy roots are capable of rapid growth as well as genetic *in vitro* stability, resulting in a greater amount of metabolites compared to non-transformed plant roots (Tindamanyire, 2014); further increases were evident with the use of elicitors (Pistelli et al., 2010). The genetic engineering tools may further instigate the production of these phytochemicals by enhancing the understanding of biosynthetic pathways. For instance, hyoscyamine is naturally abundant in *Atropa belladonna* (a drug), which retards the activity of acetylcholine in muscles and the nervous system. However, scopolamine, a medication used to treat motion sickness, is 10 times more popular than hyoscyamine. Micropropagation is a beneficial method by which the elite varieties and medicinal plants are multiplied. Multiple copies of a single cell are produced through micropropagation. An explant (plant portion) with a constructed meristem is utilized to preserve genetic integrity. The longer it takes for the plants to establish themselves in culture, the more likely they will be virus-free, especially with smaller explants. To confirm that the plantlets produced from the tips of meristems are actually virus-free, serology tests are utilized. These plants can be micro propagated to boost yield and ensure that the next crop is made of virus-free components (Taskn et al., 2013). Additionally, George et al.

(2008), Pinto et al. (2013), and Mitchell and Ahmad (2003) reported these micro propagated plants as a significant source of phytochemicals.

1.3.1.1 APPLICATIONS OF GREEN BIOTECHNOLOGY

1.3.1.1.1 Genetically modified or transgenic organisms

By transferring genes between two different organisms, modern technology allows the agricultural system to create new species while cultivating plants with commercial and technological value (Silveira et al., 2005). The first GM plants were created in 1983, and China approved the practice in 1990 (Guerrante, 2003). The first GM plants for large-scale commercial use in the U.S. were introduced in 1992 with the production of Flavr Savr tomatoes, followed by Roundup Ready soybeans in 1996 (Silveira et al., 2005). Herbicide tolerance, incredible productivity, and insect resistance to unfavorable climatic and soil conditions are all traits of GM plants (James, 2014). For instance, the *Agrobacterium tumefaciens* bacteria's glyphosate resistance gene was introduced into the DNA of soybeans to create the herbicide-tolerant GM crop known as "Roundup Ready" (Silveira et al., 2005). The glyphosate component of this herbicide kills the weeds and inhibits the manufacture of vital amino acids involved in growth for weed survival, which is dependent on the shikimate-3-phosphate synthase enzyme, as reported by Krimsky and Wrubel (1996) and Daniele et al. (1997). The genes from the bacterium type *Bacillus thuringiensis*, which encodes insecticidal action, are found in insect-resistant crops, including Bt maize and Bt cotton (James, 2014; Silveira et al., 2005). These genes produce a derived protein that demonstrates resistance towards particular species of *Lepidoptera* (Armstrong et al., 1995; Huang et al., 2002). The primary goal of these products is to control crop productivity while lowering production costs and battling pests for the benefit of farmers (Huang et al., 2002; Silveira et al., 2005). From the year 1996 to 2015, the use of GM plants increased noticeably by 100 times, from 1.7 to 179.7 M ha (million hectares). The most quickly adaptable technique in contemporary agricultural history is thought to be GM technology (James, 2014). Soy, corn, canola, and cotton are the GM crops that are produced most commonly worldwide (James, 2014). Additionally, it reduced the usage of pesticides by 37%, raised yield by 22%, increased farmers' revenue by 68%, and increased crop production by about 133.3 billion U.S. dollars between 1996 and 2013 (James, 2014). These effective gains revealed the capacities of nations that export food in the agricultural markets (Gomes and Bor'em, 2013). It was predicted that new GM crops would also replace already accessible GM varieties of crops. The future perspective of this technology aimed to increase drought and salinity stress resistance, as well as produce virus- and pesticide-free crops that will enable farmers to improve not only crop yield but also the nutrition and quality of food (James, 2014).

1.3.1.1.2 Green biotechnology and power generation tools

The green biotechnology develops plants with better production and also focuses on adaptive plants with greater potential for energy and biofuel production (Beer et al., 2009). There are many examples, such as improvements in sugarcane aimed at achieving high production, cellular fibers, sucrose content, drought resistance, and ethanol production by reducing ecological gaps (Morais et al., 2015). Additionally, it encourages the early growth and development of plants suitable for various harvesting techniques, particularly grains with higher oil content (Leite, 2015). Rhizobacteria that promote plant growth are also considered a biocontrol method for plant infections since they outperform pesticides in preventing crop damage and fostering plant growth (Tabassum et al., 2017). Furthermore, in their efforts to improve first- and second-generation biofuels, scientists have concentrated on creating renewable energy from a variety of biological sources, including microorganism-based biofuels (Machado, 2013; Mutanda et al., 2011).

1.3.1.1.3 Biopolymers production

The origins of biopolymers may come from plants, including exudates from trees, seeds from locusts and guar gum, and algae (agar, alginate, and carrageenan). Some microbial biosynthetic products such as gum, gellan, dextran, curdlan, and xanthan, as well as chemically modified natural polysaccharides such as starch, gelatin, pectin, and sodium carboxymethyl cellulose (Yu et al., 2006), are also included. Green biotechnology plays the most effective role in the synthesis of biopolymers and plant breeding. The starch obtained from potato and corn-like vegetables may be modified in order to generate raw materials with improved stability for storage purposes and less sensitivity to humidity variations, which may tend to degrade effectively (Mali et al., 2010). The main vital biopolymers are starch, xanthan, polylactide, and polyhydroxyalkanoates (PHAs) (Pradella and Nacional, 2015). The alternate focus of obtaining biopolymers instead of food crops may be shifting towards non-food crops such as straw and bagasse by using biotechnological processes. The other features such as durability, transparency, flexibility, and heat resistance are also being enhanced significantly along with the growing varieties of biopolymer (Bioplastics, 2016). The biopolymers conserve 50% higher selling market value than the commonly available biopolymers (Borschiver et al., 2008). However, the market value is anticipated to decline by 20% to 25% during the following five years, and as a result, the cost of polymers made from oil may rise. Additionally, when oil demand rises, its price may also tend to rise (Pereira et al., 2013). The main cause of the rising demand for food and other supplies is the population's rapid growth. To fulfill the requirements for food, agriculture is the major source that provides the necessary raw materials for industrial use. The products of utmost necessity may be increased with green biotechnology by improving safety and quality against pests while alleviating environmental impacts. This can be achieved by preparing raw materials of better quality and reducing the cost and time of operations.

1.3.2 RED BIOTECHNOLOGY (MEDICAL)

To combat severe life-threatening diseases and minimize their rates, medical biotechnology has involved the genetic makeup of organisms over the years. As of today, about 250 vaccines and healthcare products are available to treat untreatable diseases in patients with the help of biotechnology (Biotechnology Industry Organization, 2010). The new insights and understanding of disease-causing agents have led to diagnostic tools, therapies, and drugs that are used for the treatment of these diseases. For instance, the diagnosis of genetic diseases, such as cystic fibrosis, which is passed from one generation to the next. Moreover, cancerous diseases, such as ovarian and breast cancer, caused by genetic mutations, can be treated by testing DNA (Beroukhim, 2010; National Cancer Institute, 2009).

The molecular causes of diseases are elucidated by the biotechnology of diabetes, cancer, and rheumatic diseases. To treat the disease more accurately, scientists have been able to diagnose the genetic dispositions. One of these is the patient-specific breast cancer medication Herceptin, which was developed and is used as a personalized and individualized treatment (Roman, 2005).

1.3.2.1 APPLICATION OF RED BIOTECHNOLOGY

1.3.2.1.1 Clinical trials and research

Red biotechnology is associated with the science of clinical trials and research, which provides the impact of scientific analysis of the benefits, effects, and risks of medicines. The designed trial methodologies are adapted and conducted in clinical research.

1.3.2.1.2 Gene therapy

It is concerned with inserting desired genes into cells to treat various ailments. Gene therapy primarily entails changing an organism's genetic makeup to treat or prevent disease in humans or animals. As a result, red biotechnology also reflects the advancements in contemporary research that, through gene therapy, aid in understanding heritable disorders and their genetic foundation.

1.3.2.1.3 Genetic engineering

This technology is the operational tool that is used in the modification of genetic materials, such as DNA. Recombinant technology, also known as rDNA technology, gene alterations, and cloning form the foundation of this technology. These methods make up the bulk of genetic engineering.

1.3.2.1.4 Gene diagnoses

For the study of an entire gene, the detection of the gene is a necessary step, and after that, there is a need to diagnose and isolate the target genes.

The PCR technique is employed for gene diagnosis. For diagnosing genes, the polymerase chain reaction (PCR) method is frequently used. It is a thermocycler that is used to amplify certain DNA for further study.

1.3.2.1.5 The recombinant vaccine production

The new generation vaccines are mainly produced by recombinant DNA technology. The vaccinations are classified into three main groups: vector recombinant, subunit recombinant, and attenuated recombinant vaccines. There is a chance to vaccinate against a number of diseases by using one recombinant vaccinia virus.

1.3.2.1.6 Cancer and treatment

Red biotechnology is crucial to the study and therapy of cancer. Cancer differential diagnosis, early detection, cancer progression control, and effective cancer treatment are all stated benefits of biotechnology. Many biotechnology-based methods are employed in various cancer research programs.

1.3.2.1.7 Stem cells

Stem cells are used to transplant stem cells into the body's damaged areas. Stem cells are mostly used in bone marrow transplants, the repair of injured heart and nerve tissues, and bone marrow regeneration. Modern medical science greatly benefits from stem cell technologies.

1.3.2.1.8 Disease diagnosis

There are numerous diseases that are caused by genetic defects or pathogens. The diagnosis and identification of such diseases is based upon the study of genetic material of living organism deoxyribonucleic acid (DNA). The individual gene or a set of genes in an organism is identified for the detection of disease-causing pathogens.

1.3.2.1.9 Tissue engineering

It is the process of growing human tissues outside the body for later use in replacement. In modern medical science, the tissue culture technique is known as a vital strategy against various diseases.

1.3.2.1.10 Biopharmaceuticals

Biotechnology and pharmaceutical science are combined to create biopharmaceuticals. To prevent and treat the condition, byproducts and components of living organisms are utilized.

1.3.2.1.11 Medicines, new drugs and antibiotics

Medical biotechnology is linked with the widespread use of treatments, prevention, and diagnosis of different diseases. Numerous antibiotics are used to treat bacterial-related infections. Utilizing biotechnology-based methodologies, procedures, and novel drug discoveries for the development of enhanced infection and illness treatments may increase the quality of medicine.

1.3.2.1.12 Manipulation of gene expression in host cells

For the purpose of attaining the maximum expression of cloned genes, the process of manipulation is adapted. The molecular influences regulate gene expression via RNA translation or DNA transcription.

1.3.3 BLUE BIOTECHNOLOGY (MARINE)

Biotechnological tools for the benefit of society from freshwater and marine resources are associated with blue biotechnology (Foresight Marine Panel, 2005). Medicine, healthcare, nutrient-rich food supplements, and cosmetics are the main products of blue biotechnology. Additionally, bioremediation and biofuels are the industrial and environmental applications of this technology. Blue biotechnology is known by its sources, not by its market values (Kim, 2012). Furthermore, a variety of medicinal compounds are abundant in the aquatic environment. The promising use of marine biotechnology for human disease treatment is drug discovery. There are now substances with antioxidant, anticancer, antiviral, anticoagulant, antibacterial, and anti-inflammatory properties, thanks to the many marine-derived bioactive medicines (Malaker and Ahmad, 2013). Many marine creatures serve as significant sources of new secondary metabolites (Thakur et al., 2005). The pharmaceutical business based on marine biotechnology has acknowledged marine environments as a significant beneficial resource for research in the medical field on a global scale (Mayer et al., 2010). Seaweeds, accompanied by their components: (i) crab shells derived chitin and chitosans; (ii) lipids and natural colorants; and (iii) microalgae-derived vitamins, are the three primary categories of marine raw material (Polin, 2012).

The other compounds derived from marine biotechnology are enzymes and DNA polymers that are useful in pharmaceutics. They are used in PCR reactions and also have uses in the diagnosis and discovery of genes (Marsic et al., 2008).

Dairy products and processed meats comprise polysaccharides from red and brown algae, along with various species of seaweeds, which are being used as stabilizing or thickening agents. A protein called fish collagen has been utilized as a gelatin alternative. Many nutraceutical products, including fish oil capsules, contain omega-3 fatty acids. Marine algae have been used as a source for many food coloring ingredients (Rasmussen and Morrissey, 2007). Due to their thermal and elastic properties, biodegradable bioplastics like PHA are produced by a variety of microorganisms, some of which are of marine origin (Xiao and Jiao, 2011). Due to their environmental friendliness, PHA is frequently used to replace plastic made from petroleum.

1.3.3.1 PRODUCTION OF BIOACTIVE COMPOUNDS

A vast variety of chemicals obtained from marine species are being investigated through marine biotechnology, and pharmaceutical companies are producing these medications. The crucial components among them are the polyunsaturated fatty acids, enzymes, proteins, and carotenoids (De Morais and De Morais, 2013; Murray et al., 2013). All of these compounds could be used in the pharmaceutical sector as antibacterial, antifungal, analgesic, anti-inflammatory, and antioxidant factors (Murray et al., 2013). The marine species that can be exploited to produce bioactive compounds include fungi, sponges, marine yeast, and micro- and macroalgae. These substances can be created from cultures or by genetic engineering by employing photobioreactors in controlled environments (Rorrer and Cheney, 2004). Besides that, different methods such as enzymatic hydrolysis, solvent extraction, and mechanical pressure are applied to extract these compounds (Murray et al., 2013).

1.3.3.2 APPLICATIONS IN FOOD INDUSTRY

The marine biotechnology contributes to the food industry with the production of pigments and additives obtained from marine microalgae and cyanobacteria. Marine microalgae have the capability to produce polyunsaturated fatty acids (Batista et al., 2014), astaxanthin (Tripathi et al., 1999), lutein (Sanchez et al., 2008), and β-carotene (Qin et al., 2008). Pulz and Gross (2004) reported cyanobacteria as the producers of phycocyanin. The conditions of the processes and various applied techniques can vary from case to case. Biomass production and its applications are the main objectives for producing food from marine microalgae. Main applications include carbohydrates, nutritional supplementation, and pigments (Becker, 2004). Mostly, biomass production is employed in powder form for food applications. Moreover, it has significant importance for the health of consumers in the form of mineral fortification and supplementation. The powdered form of microalgae has been used in yogurt, pasta, and bread in many countries like China, the USA, Japan, France, and Germany (Pulz and Gross, 2004). Conversely, there are many factors determining algal biomass production, such as temperature, pH, nutrients, light, and agitation, that affect biomass production (Miao and Wu, 2004).

1.3.3.3 COSMETIC INDUSTRY

Marine biotechnology presents some new compounds that play a key role in the cosmetics industry for the manufacturing of valuable cosmetic products (Chisvert and Salvador, 2007). The marine compounds are obtained and incorporated biotechnologically into some cosmetics that are already being marketed. For instance, the *Micrococcus* lysate enzymatic extract contains the DG-DNA complex R as one of its components. The *Micrococcus*, obtained from oceans and soils, has a higher UV light tolerance. Additionally, it contains the UV-endonuclease enzyme, which speeds up the healing of sun damage. Currently, this substance is utilized in a variety of face creams (Emiliano et al., 2012). Compounds used in the creation of cosmetics are produced by isolating them from microalgae or other marine creatures, then cultivating them in photobioreactors (Biotech marine, 2016). Although only a few scientific studies are available in this field, many companies are investing in cosmetic products obtained from marine biotechnology. These efforts should be encouraged to improve the techniques involved in the production and application of such compounds that are synthesized from marine organisms.

1.3.3.4 MARINE RESOURCES AND BIOFUELS

The main sources of biofuels are marine macro- and microalgae that developed as a substitute for using land plants (Cunha, 2012; Hargreaves, 2012). They neither compete with agricultural land nor need chemical fertilizers because they have a marine origin, taking advantage of being easily grown in water (Hargreaves, 2013). Additionally, they secure a higher production rate, as reported by Georgianna and Mayfield (2012), and do not require raw materials based on food (Defanti et al., 2010). On the other hand, the massive yield and high costs of the processes—filtration, fermentation, transesterification, extraction, and centrifugation—pose difficulties for the commercial viability of marine-origin biofuels (Mota and Monteiro, 2013). Bioethanol, sometimes referred to as 3G ethanol, is one of the most significant biofuels created by marine biotechnology. It is a product of the fermentation of algal biomass and continues to have a more prominent role than other algal-produced biofuels (Goh and Lee, 2010). The potential and growing significance of using blue biotechnology in various industries, such as pharmaceuticals, food, biofuels, and cosmetics, is evident. Marine biotechnology has produced positive results and shown that its use is feasible despite being a young field.

1.3.4 WHITE OR INDUSTRIAL BIOTECHNOLOGY

It is linked with the industrial application of technology in which renewable sources are utilized to produce the end products and intermediates. Enzymes and microorganisms are used to produce useful industrial products. It started in the 1800s when Louis Pasteur proved that microbial activity was involved in the process of fermentation. After that, Fleming discovered molds as antibacterial compounds, and Florey and Chain improved it as penicillin in the early 1900s (Raju, 1999). Since then, the use of several contemporary technologies has revolutionized human life. Human proteins made by microbes like *E. coli* and *S. cerevisiae*, such as growth-related hormones and insulin, are used in this technology. Interferon (clot-dissolving protein), also known as tissue plasminogen activator, and other therapeutic proteins made in yeast and bacteria are increasingly utilized to treat some malignancies and skin conditions (Ponstein, 1996).

1.3.4.1 RECOMBINANT PROTEINS FROM PLANTS

A soil bacterium performs the plant transformation. The use of *A. tumefaciens* is common. Using bacteria, the target genes are translated into plant cells. Crown gall disease is a plant pathogen caused by *A. tumefaciens* that develops tumors in vulnerable plants. The tumor-inducing Ti (tumor-inducing) plasmid found in the bacterium, after inserting its T-DNA into the cell, results in tumor development. To insert desired genes into the plant, modified Ti plasmids along with deleted virulence genes are employed (Gelvin, 2003). This approach to genetic engineering is being investigated and might be used to enhance the content of specific metabolites in medicinal plants. Secondary metabolite production via plant cell culture has become a significant topic of study involving numerous scientific fields, including pharmacognosy (Khan et al., 2009). Transgenic therapeutic plants are being created using a variety of biotechnological procedures. Plants like the alkaloid-producing *Hyoscyamus muticus* and *Arabidopsis thaliana* are being transformed using different biotechnological tools. Various desired agronomic qualities are imparted to these altered plants, or large concentrations of vaccine antigens and other biopharmaceuticals are expressed in the plants (De La Riva et al., 1998). New clinical medications like antibiotics and reserpine (the first tranquilizer), which were made from natural sources, are expected to ultimately be synthesized (Phillipson, 2007).

1.3.4.2 MAIN APPLICATIONS OF WHITE BIOTECHNOLOGY

1.3.4.2.1 Chemical ingredients production

The white biotechnology contributes to the reduction of gas emissions and energy savings through the bulk chemical production of alcohols and organic acids (Hermann et al., 2007). This is an important characteristic that is involved in the development of various large-scale processes (Villadsen, 2007). The main food additives, such as lactic acid, citric acid, and acetic acid, are now prepared on an industrial scale by applying white biotechnology. *Aspergillus niger* produces citric acid, a commercially significant substance with a market value of up to 1.4 billion U.S. dollars (Carvalho et al., 2007). Sweeteners are another important class, including xylitol, aspartame, and sorbitol. The demand for these products, due to their zero calories, is increasing globally among consumers in recent years. New metabolic pathways are also being adapted to synthesize novel products by combining biotechnology and synthetic biology (Philippe et al., 2014). Table 1.1 enlists these chemicals' biotechnological production along with production scale and substrate information.

Some of the other primary products, such as vitamins and amino acids, are produced completely or partially by utilizing this biotechnology. Because they contain proteins that play a significant role in some metabolic processes, amino acid molecules are important for both humans and animals. There are 20 common amino acids in total, with some being essential (Mitsuhashi, 2014). Table 1.2 lists the vital primary amino acids created by white biotechnology, followed by substrate and production size.

1.3.4.3 OTHER INDUSTRIAL COMPOUNDS

Surfactants, enzymes, and flavors are a few of the additional substances produced by white biotechnology that have significant economic importance. Based on industrial processes and microorganisms, enzymes including lipases, proteases, and oxidases are produced. The anticipated value of the global enzymes market was 2.9 billion U.S. dollars in 2008 and is expected to reach 4.4 billion U.S. dollars in 2015 (Erickson and Winters, 2012). Various industries, including the food, pharmaceutical, and cosmetic industries, exhibit aroma compounds as an important class (Forti et al., 2015). White biotechnology provides an alternative to chemical procedures and natural extraction for the synthesis of aromas, and it is carried out under gentle conditions with excellent specificity (Molina, 2010). The market value was predicted to be 24.7 billion U.S. dollars in 2016, and continued growth is anticipated over the ensuing years (Leffigwell, 2016).

TABLE 1.1 The Vital Chemical Products Along with Starting Material and Production Scale by Adapting White Biotechnology

	Biotechnological Production	Substrate	Production Scale	Applications	References
Ethanol	Fermentation contributes total 90% of production of the world with estimated yield of 90% to 95%	Sugarcane and corn kernel biomass	1 g/L production rate. 86.1 × 106 m^3 world production	Food and fuel industry, chemical sectors and as fuel additives	Lennartsson et al. (2014); Patel (2006)
Acetic Acid	Fermentation and methanol carboxylation synthesis in food industry	Ethanol	190.000 tons/year World production	Food and chemical industries	Carvalho et al. (2005)
Lactic Acid	Yield greater than 90% by fermentation	Yeast extract	771 g/L production rate. 390 tones/Y world production rate	Beverage, food industries, personal, and care-products	Carvalho et al. (2005); Guilherme et al. (2009)
Glycerol	By-product (fat and oil processing), and also produced by fermentation and low cost U.S. $1.000/ton.	Oils and fats	–	In making cheap oils, multiple applications in chemical industry	Patel (2006)
3-Hydroxypropanoic acid	Fermentation of glucose with micro-organisms (modified)	Glycerol and glucose	–	Numerous chemicals of industrial importance	Patel (2006)
1,3-propanodiol	Fermentation by utilizing GM microorganisms with 51%–55% yields	Glycerol and glucose	–	As vital biopolymers	Patel (2006)
Acrylic Acid	Developed biotechnologically along with GM micro-organisms	–	–	As biopolymers	Patel (2006)
C_4 Succinic Acid	Carbohydrates fermenting process	Sugars	–	Industries (Food, agriculture, medicine)	Erickson and Winters (2012); Patel (2006)
Fumaric Acid	Carbohydrate fermentation at small scale	Sugars	Estimated production of 12.0 ton/y	In beverages as ingredients, food, as additive in animal food	Patel (2006)

TABLE 1.2 Biotechnologically Obtained Amino Acids Products Along with Substrate and Production-Scale by Utilizing White Biotechnology

	Biotechnological Production	**Substrate**	**Production Scale**	**References**
Leucine, valine and isoleucine	Obtained from *Corynebacterium Glutamicum* (GM)	–	–	Becker and Wittmann (2012)
Lysine	Obtained from *Corynebacterium glutamicum* (GM)	Biomass of lignocellulose or glucose	1.30 M tons/y	Mitsuhashi (2014)
Threonine	Produced by *E. coli*	–	Estimated production of 70.0 tons	Leuchtenberger et al. (2005)
Methionine	Obtained from *Corynebacterium glutamicum* genetically modified with *E. coli* or *C. lilium*	Ammonium sulfate and glucose	3 g/L Mean yield, 400 tons estimated production with 2.20 billion U.S.$ (market value)	Kromer et al. (2006); Kumar et al. (2005); Mitsuhashi (2014)
Histidine	Same substrate for lysine	–	Yield of 8.0 g/L	Mitsuhashi (2014)
Phenylalanine	Obtained from fermentation with *E. coli*	Glucose	Production of 25.0 g/L	Leuchtenberger et al. (2005)
Tryptophan	Obtained from *E. coli*	Glucose	3.0 tons estimated yield	Leuchtenberger et al. (2005)
Glutamate	Same for lysine	Carbohydrates	2.0 M tons/y and, yield of until 100 g/L	Becker and Wittmann (2012); Becker et al. (2011)

1.3.4.4 SUSTAINABLE OR GREEN CHEMISTRY

Numerous various types of compounds, including bulk chemicals, fine chemicals, polymers, fuels, and medicinal products, are produced in the industrial zone. The development of sustainable chemical processes that enhance human well-being and safeguard the environment while maintaining a reasonable commercial value is one of chemists' greatest difficulties. The ultimate objective is to create clean chemical technology using waste-free renewable energy and raw materials, with maximum productivity and competitiveness. Sustainable chemistry is based on a variety of technologies, including improved catalysts, recycling and reuse technology, novel separation techniques including membrane processes, and more efficient conventional chemical processes. Due to the fact that biotechnology is inherently suited for sustainable chemistry, the latter technology is having an increasing impact on the chemical industry (Eggersdorfer et al., 1992; Gavrilescu and Chisti, 2005; Wilke and Vorlop, 2004) (Figure 1.1).

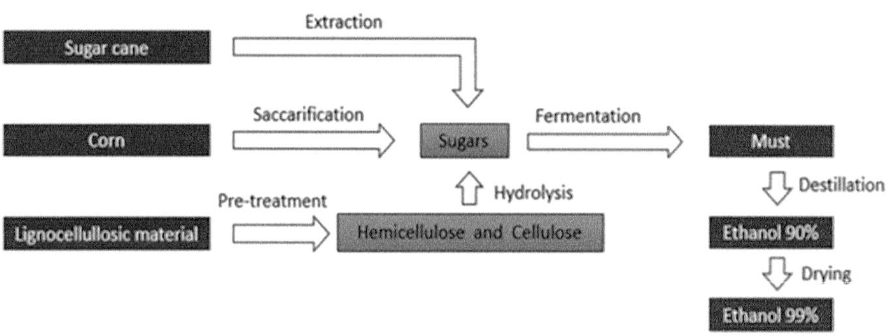

FIGURE 1.1 The ethanol production from principal raw materials obtained from plants.
Source: Adapted from Soccol et al. (2009).

1.3.4.5 INDUSTRIAL BIOTECHNOLOGY-BASED PROCESSES

1.3.4.5.1 Fermentation

A vast range of chemicals, including vitamins, solvents, citric acid, antibiotics, amino acids, industrial enzymes, steroids, and many others, are produced by using industrial biotechnology. The primary technology used in this case is industrial fermentation, which involves the fermentation of bacteria, yeasts, and fungi that effectively transform carbohydrates into valuable compounds. For several of these products, it is the only technique of industrial manufacturing, and some are produced in extremely large amounts (Baets et al., 2000; Demain, 2007; Goldberg et al., 2006). The selection includes everything from cheap bulk supplies to extremely pricey fine chemicals.

1.3.4.5.2 Enzymatic technology and processing

Proteins that are catalytically active and have undergone the evolution of billions of years are known as enzymes. They serve as extremely effective catalysts that control the life chemistry without the need for the high pressures, intense temperatures, and corrosive environments that are frequently necessary in chemical synthesis processes. They are the biological world's machinery and function as more of their extraordinary capabilities. The name of this technical field is biocatalysis (Vandamme et al., 2006). For biocatalytic reactions, enzymes have grown to be crucial in a variety of industrial industries. Typically, microbial enzymes from the aforementioned fermentation methods are employed. Amylase, a very thermostable enzyme that hydrolyzes starch at 105°C, is a crucial enzyme. These thermostable enzymes significantly speed up bioreactions by enabling them to occur at high temperatures. One such significant enzyme in this field is glucose isomerase. Fructose is produced by this enzyme from glucose.

1.3.4.5.3 Biotechnology and socio-economic advantages

When biotechnological process steps are incorporated into chemical synthesis, there are frequently significant ecological benefits, including significantly reduced waste generation, lower energy needs, less need for solvents, removal of hazardous intermediate products, and so forth. The move in technology is often not motivated by these ecological benefits, though. The environmental benefits are a welcome byproduct of such a decision and, by themselves, are insufficient to persuade decision-makers to implement new technology. Instead, it is almost always the technological process improvements and associated cost reductions that are the driving factor. Industrial biotechnology is frequently linked with eco-ecological advancements: the enhanced productivity and lower production costs of biotechnological processes mostly result in a significant reduction in ecological impact and typically enhance competitiveness. Around 21 case studies were provided in an OECD report from 2001 (Griffiths, 2001), each of which strongly demonstrated the benefits of industrial biotechnology in terms of both the economy and the environment. In many cases, the applied techniques are competitive in industrial practices; they are not, in any way, restricted to only research and theoretical studies.

1.3.5 YELLOW BIOTECHNOLOGY

Yellow biotechnology is the area of biotechnology research that is most popular worldwide, followed by red, green, and white biotechnology. Insects need to be extensively researched in order to be used in agriculture and industry to produce nutrient-dense food for everyone. Thus, using them can be a solution to the problem of meeting the growing need for food without endangering the environment. The oldest subset of biotechnology is known as yellow biotechnology. Nutritional biotechnology, along with insect biotechnology, are alternative names for yellow biotechnology since it uses insects or microbes to produce food that is more nutrient-dense for us. On the other hand, it involves certain methods through which products are produced on an industrial scale by insects in the areas of agriculture and healthcare. However, traditional food products are poor in nutrients, and these nutrients are enhanced by adapting techniques such as *in-vitro* cell culture techniques and genetic modifications. The microbial or enzymatic activities are managed to enhance the nutrients of food crops with health-friendly additives.

1.3.5.1 DEMAND FOR YELLOW BIOTECHNOLOGY

Crop cultivation and meat production are two examples of traditional food production. Because crop agriculture requires a lot of land, it encourages the destruction of forests. Additionally, it requires fresh water, fertilizers, pesticides, and other chemicals that directly contaminate land and groundwater. Every year, insects cause a significant portion of the crop to be destroyed, and the use of chemical insecticides damages the environment. Large tracts of land, pristine water, rich feed, antibiotics, vaccinations, hormones, and so forth are also necessary for the production of meat. Land and water are also contaminated by the chemicals and animal waste used in the meat industry. The meat sector is a major source of the greenhouse gases that cause global warming. It is now difficult for scientists to meet the rising food demand without endangering the ecosystem.

1.3.5.2 MAJOR GOALS OF YELLOW BIOTECHNOLOGY

Yellow biotechnology's main objective is to provide people with high-nutrient, high-quality meals that include additions beneficial for their health without endangering (or polluting) the environment. Yellow biotechnology employs a variety of methods, including enzymatic, micro-biological, and molecular biology genetics approaches, to achieve its objectives. By utilizing biotechnological methods, yellow biotechnology exploits insects in order to advance the bioeconomy.

1.3.5.3 YELLOW BIOTECHNOLOGY AND INSECTS

A few decades ago, insects were viewed negatively by humans because they destroyed our crops, vied for our food supplies, and spread numerous diseases. However, insects are now incredibly beneficial to us thanks to applied entomology and biotechnological approaches. Active chemicals for use in agriculture, medicine, and industry are currently produced using insects. Chitinases, cellulases, and insect antenna-based enzymes all originate from insects. The most prevalent use of insects in food or nutritional engineering is in biosensors. Insect biotechnology has a wide range of possible research areas, such as antimicrobial peptides (AMPs), the red flour beetle model to examine gene-food ingredient interactions, and the fruit fly *Drosophila* to investigate aging, metabolic diseases, and cancer.

1.3.5.4 YELLOW BIOTECHNOLOGY APPLICATIONS

We can learn a lot from them, such as how they naturally digest different kinds of food sources and how they protect themselves against different diseases and adversaries. We can investigate new sources of nutrition and goods that can be used for human well-being by employing the active chemicals and enzymes that they use to protect themselves. The following categories cover the main uses of yellow biotechnology.

1.3.5.5 INSECT DERIVED ENZYMES FOR WHITE BIOTECHNOLOGY

Insects have specialized enzymes that allow them to consume any organic compounds. These specialized enzymes can be applied to the organic waste management business and the food industry. In the food and fermentation industries, enzymes generated from insects are frequently employed for a variety of processes, including saccharification, gluten degradation, cellulolytic enzymes, and so forth. The primary enzymes obtained from insects include peptidases, lipases, amylases, and β-D-glucosidases.

1.3.5.6 NUTRITIONAL FOOD PRODUCTION

Generating high-quality food, enhancing it, and providing nutrients using fortification, enzymatic, microbiological, and GMO technologies. Using fermentation technology to improve the flavor, nutrition, and scent of food. Storage, protection from numerous hazardous bacteria, and improvement of shelf life for food.

1.3.5.7 GENETICALLY MODIFIED (GM) CROPS

The primary method for changing specific gene codes in organisms, plants, or animals is genetic engineering. Therefore, more nutrient-dense food is produced by eliminating undesirable genes from plants or animals and introducing beneficial ones. Most commonly planted agricultural crops, including maize, rice, wheat, paddy, tomato, eggplant, beans, cauliflower, potatoes, apples, and oranges, have already undergone genetic modification and have been commercially successful.

1.3.5.8 IN-VITRO CELL CULTURE TECHNOLOGY

Test-tube meat was also produced using techniques for cultivating animal tissue. Adult cow stem cells are employed in *in vitro* cell culture technologies to produce meat.

1.3.5.9 DRUGS FROM INSECTS

The common green bottle fly, *Lucilia sericata*, is used to treat chronic or non-healing wounds. It was discovered that *in vitro* larval secretions of *Lucilia sericata* improved wound closure.

1.3.6 GREY BIOTECHNOLOGY

Grey biotechnology is synonymous with green biotechnology, which emphasizes the use of living organisms for the improvement of the environment. Microorganisms are used to remove pollutants from the environment. It is also beneficial for protecting flora and fauna. Human waste management and disposal are carried out using microorganisms. An example of this technology is the removal of plastic from the environment, as plastic is dangerous for the environment and human health and is non-biodegradable.

1.3.7 BROWN BIOTECHNOLOGY

The grey biotechnology and brown biotechnology are similar to each other. Brown biotechnology is also sometimes referred to as desert biotechnology. The purpose of this biotechnology is to make the soil of deserts more productive. To attain this goal, drought-resistant varieties were introduced for positive use. Better-quality seeds are prepared using GM technology that can grow in areas where water availability is very low. There are genetically engineered crops that already exist and are resistant to such desert conditions. Drought Gard is an example of such a crop, which ensures the survival and better yield of corn plants under poor water conditions. About 70% of the Earth's land is dryland, and 1 billion people are affected by drought. In order to solve the issue of water scarcity and meet the demand for food to end hunger, GM crops like alfalfa and sweet potatoes are being developed for this purpose.

1.3.8 VIOLET BIOTECHNOLOGY

It is a unique technique that deals with the study of the legal aspects of biotechnology. It is based on the study of moral and ethical values and philosophical aspects. It also studies the moral values regarding biosecurity and gene therapy. It was created in 1980 after the verdict of the Supreme Court in the U.S. The patented use of genetically altered microorganisms is possible, according to Chemic. Invention patent rights, research publication, and protecting intellectual property rights are also included in this technology.

1.4 CONCLUSION

Biotechnology deals with the issues of food and health security around the globe. The application of biotechnology has led to a better life by controlling plant and animal diseases and enhancing the yield of food crops with improved health. The different branches of biotechnology focus on the betterment of individuals while keeping the environment safe and adapting eco-friendly tools. The major industrial-based products of greater importance have revolutionized human life with massive economic and social benefits. GM crops, the drug industry, disease diagnosis, eco-friendly pests, and the food industry are the major applicable fields of modern biotechnology. It can be concluded that biotechnology, in the near future, will override all other technologies with its products that will be entirely available in each field of demand.

KEYWORDS

- **biotechnology**
- **blue biotechnology**
- **genetically modified (GM) crops**
- **green biotechnology**
- **industrial applications**
- **red biotechnology**
- **white biotechnology**

REFERENCES

Abdin, M. Z., Zhu, Y., Tan, B., Bay, B., & Liu, C. (2007). Enhancing bioactive molecules in medicinal plants. In *Natural Products—Essential Resources for Human* (pp. 45–57). World Scientific Publishing Co. Pvt. Ltd.

Amarakoon, C. L., Hamilton, S. A., Mitchell, P. F., Tennant, M. E., & Roye. (2017). *University of the West Indies, Mona, Kingston 7, Jamaica.*

Batista, I., Martins, G., Padilha, M., do Castelo Paulo, M., & Bandarra, N. M. (2016). Production of squalene and polyunsaturated fatty acids by microorganisms of the Thraustochytrids group. *Biotechnology, 201*, 444.

Becker, J., & Wittmann, C. (2012). Systems and synthetic metabolic engineering for amino acid production—the heartbeat of industrial strain development. *Current Opinion in Biotechnology, 23*(5), 718–726.

Becker, J., Zelder, O., Häfner, S., Schröder, H., & Wittmann, C. (2011). From zero to hero—Design-based systems metabolic engineering of *Corynebacterium glutamicum* for *L*-lysine production. *Metabolic Engineering, 13*(2), 159–168.

Becker, W. (2004). Microalgae in human and animal nutrition. In A. Richmond (Ed.), *Handbook of Microalgal Culture: Biotechnology and Applied Phycology*.

Beer, L. L., Boyd, E. S., Peters, J. W., & Posewitz, M. C. (2009). Engineering algae for biohydrogen and biofuel production. *Current Opinion in Biotechnology, 20*(3), 264–271.

Behera, S. S., Ray, R. C., Das, U., Panda, S. K., & Saranraj, P. (2019). Microorganisms in fermentation. In *Essential Fermentation Technology* (Vol. 1, pp. 1–39).

Beroukhim, R., Mermel, C. H., Porter, D., Wei, G., Raychaudhuri, S., Donovan, J., & Meyerson, M. (2010). The landscape of somatic copy-number alteration across human cancers. *Nature, 463*(7283), 899–905.

Bioplastics, E. (2016). *Market Report*. Published by European Bioplastics in collaboration with the Nova-Institute.

BiotechMarine. (2016). Company profile and marine biotechnology innovations.

Biotechnology Industry Organization. (2010). *Healing, Fueling, Feeding: How Biotechnology is Enriching Your Life.* Retrieved from: http://www.ValueofBiotech.com (accessed on 18 March 2025).

Borschiver, S., Almeida, L. F., & Roitman, T. (2008). Monitoramento tecnológico e mercadológico de biopolímeros. *Polímeros, 18*, 256–261.

Buchholz, K., & Collins, J. (2013). The roots—A short history of industrial microbiology and biotechnology. *Applied Microbiology and Biotechnology, 97*, 3747–3762.

Bud, R. (1994). *The Uses of Life: A History of Biotechnology*. Cambridge University Press.

C. elegans Sequencing Consortium. (1998). Genome sequencing of the nematode *C. elegans*: A platform for investigating biology. *Science, 282*, 2012–2018.

Carvalho, W., Silva, D. D. V., Canilha, L., & Mancilha, I. M. (2005). Aditivos alimentares produzidos por via fermentativa parte I: Ácidos orgânicos. *Revista Analytica, 18*, 70–76.

Chaudhuri, K. (2013). *Recombinant DNA Technology*. The Energy and Resources Institute (TERI).

Chisvert, A., & Salvador, A. (2007). *Analysis of Cosmetic Products*. Elsevier.

Cunha, T. P. E. (2012). *Blue Growth for Portugal: Uma Visão Empresarial da Economia do Mar*. Coltec.

Cunningham, C., & Porter, A. J. R. (1998). *Recombinant Proteins from Plants: Production and Isolation of Clinically Useful Compounds.* Totowa, NJ: Humana Press.

Daniele, P. G., De Stefano, C., Prenesti, E., & Sammartano, S. (1997). Copper (II) complexes of *N*-(phosphonomethyl) glycine in aqueous solution: A thermodynamic and spectrophotometric study. *Talanta, 45*(2), 425–431.

De Baets, S., Vandedrinck, S., & Vandamme, E. (2000). Vitamins and related biofactors, microbial production. In J. Lederberg (Ed.), *Encyclopedia of Microbiology* (2nd ed., Vol. 4, pp. 837–853). Academic Press.

De La Riva, G. A., González-Cabrera, J., Vázquez-Padrón, R., & Ayra-Pardo, C. (1998). *Agrobacterium tumefaciens*: A natural tool for plant transformation. *Electronic Journal of Biotechnology, 1*(3), 24–25.

Defanti, L. S., Siqueira, N. S., & Linhares, P. C. (2010). Produção de biocombustíveis a partir de algas fotossintetizantes. *Revista de Divulgação do Projeto Universidade Petrobras e If Fluminense*, 11–21.
Demain, A. L. (2000). Small bugs, big business: The economic power of the microbe. *Biotechnology Advances, 18*(6), 499–514.
Demain, A. L. (2007). The business of biotechnology. *Industrial Biotechnology, 3*, 269–283.
Diamond, J. (2002). Evolution, consequences, and future of plant and animal domestication. *Nature, 418*(6898), 700–707.
Durate, I. (2001). *Influência do Meio Nutricional no Crescimento e Composição Centesimal de Chlorella sp (Chlorophyta, Chlorococcales)* (Doctoral dissertation). Universidade Estadual Paulista.
Emiliano, A., Guimarães, F., & Netz, D. J. A. (2012). *Biotecnologia Na Obtenção De Ativos E Excipientes Cosméticos.*
Erickson, B., & Winters, P. (2012). Perspective on opportunities in industrial biotechnology in renewable chemicals. *Biotechnology Journal, 7*(2), 176–185.
Foresight Marine Panel (FMP), Marine Biotechnology Group. (n.d.). A study into the prospects for marine biotechnology development.
Forti, L., Di Mauro, S., Cramarossa, M. R., Filippucci, S., Turchetti, B., & Buzzini, P. (2015). Non-conventional yeasts whole cells as efficient biocatalysts for the production of flavors and fragrances. *Molecules, 20*(6), 10377–10398.
Gavrilescu, M., & Chisti, Y. (2005). Biotechnology—A sustainable alternative for chemical industry. *Biotechnology Advances, 23*(7, 8), 471–499.
Gelvin, S. B. (2003). *Agrobacterium*-mediated plant transformation: The biology behind the "gene-jockeying" tool. *Microbiology and Molecular Biology Reviews, 67*(1), 16–37.
George, E. F., Hall, M. A., & De Klerk, G. J. (2007). *Plant Propagation by Tissue Culture: Volume 1. The Background* (Vol. 1). Springer Science & Business Media.
Goh, C. S., & Lee, K. T. (2010). A visionary and conceptual macroalgae-based third-generation bioethanol (TGB) biorefinery in Sabah, Malaysia as an underlay for renewable and sustainable development. *Renewable and Sustainable Energy Reviews, 14*(2), 842–848.
Goldberg, I., Rokem, J. S., & Pines, O. (2006). Organic acids: Old metabolites, new themes. *Journal of Chemical Technology and Biotechnology: International Research in Process, Environmental and Clean Technology, 81*(10), 1601–1611.
Gomes, W. S., & Borém, A. (2013). Biotechnology: New paradigm of Brazilian agribusiness. *Journal of Economics and Agribusiness, 11*(1).
Goudie, A. S. (2018). *Human Impact on the Natural Environment.* John Wiley & Sons.
Griffi, M. (2001). *The Application of Biotechnology to Industrial Sustainability* (OECD report). Retrieved from: http://www1.oecd.org (accessed on 18 March 2025).
Guerrante, R. D. S. (2003). *Transgênicos: Uma visão estratégica.* Interciência.
Guilherme, A. D. A., Pinto, G. A. S., & Rodrigues, S. (2009). Avaliação da produção de ácido lático por *Leuconostoc mesenteroides* B512F em xarope de caju. *Food Science and Technology, 29*, 738–747.
Hargreaves, P. I. (2013). *Produção De Etanol A Partir De Kappaphycus Alvarezii—Biocombustível De Terceira Geração* (Doctoral dissertation).

Hartmann, E. M., Durighello, E., Pible, O., Nogales, B., Beltrametti, F., Bosch, R., & Armengaud, J. (2014). Proteomics meets blue biotechnology: A wealth of novelties and opportunities. *Marine Genomics, 17*, 35–42.

Hermann, B. G., Blok, K., & Patel, M. K. (2007). Producing bio-based bulk chemicals using industrial biotechnology saves energy and combats climate change. *Environmental Science & Technology, 41*(22), 7915–7921.

Huang, F., Buschman, L. L., Higgins, R. A., & Li, H. (2002). Survival of Kansas Dipel-resistant European corn borer (Lepidoptera: Crambidae) on Bt and non-Bt corn hybrids. *Journal of Economic Entomology, 95*(3), 614–621.

James, C. (2014). *Global Status of Commercialized Biotech/GM Crops* (ISAAA Brief 49).

Kafarski, P. (2012). Rainbow code of biotechnology. *Chemik, 66*(8), 811–816.

Kennedy, M. J. (1992). The evolution of the word 'biotechnology'. *Trends in Food Science & Technology, 3*, 154–156.

Khan, M. Y., Aliabbas, S., Kumar, V., & Rajkumar, S. (2009). Recent advances in medicinal plant biotechnology. *Indian Journal of Biotechnology, 8*(1).

Kim, S. K. (2012). *Marine Pharmacognosy: Trends and Applications*. CRC Press.

Kingdom. (2005). *Volume 1–Strategy.* Retrieved from: http://www.biobridge.co.uk/LinkClick.aspx?fileticket5XPE70J4D-wM%3D&tabid567 (accessed on 18 March 2025).

Krimsky, S., & Wrubel, R. P. (1996). *Agricultural Biotechnology and the Environment: Science, Policy, and Social Issues* (Vol. 13). University of Illinois Press.

Krömer, J. O., Wittmann, C., Schröder, H., & Heinzle, E. (2006). Metabolic pathway analysis for rational design of L-methionine production by *Escherichia coli* and *Corynebacterium glutamicum*. *Metabolic Engineering, 8*(4), 353–369.

Kumar, D., Subramanian, K., Bisaria, V. S., Sreekrishnan, T. R., & Gomes, J. (2005). Effect of cysteine on methionine production by a regulatory mutant of *Corynebacterium lilium*. *Bioresource Technology, 96*(3), 287–294.

Leffingwell. (2016). *2011–2015 Flavor & Fragrance Industry Leaders.* Report by Leffingwell & Associates.

Leite, D. C. (2015). *Mapeamento De Qtl E Análises De Espectroscopia Para Teor De Óleo Visando Aplicação Em Programas De Melhoramento De Soja.*

Lennartsson, P. R., Erlandsson, P., & Taherzadeh, M. J. (2014). Integration of the first- and second-generation bioethanol processes and the importance of by-products. *Bioresource Technology, 165*, 3–8.

Leuchtenberger, W., Huthmacher, K., & Drauz, K. (2005). Biotechnological production of amino acids and derivatives: Current status and prospects. *Applied Microbiology and Biotechnology, 69*, 1–8.

Ling, N. R. (1983). Properties of the monoclonal antibodies produced by hybridoma technology and their application to the study of diseases. *Immunology, 49*(1), 202.

Machado, C. M. M. (2013). *Microrganismos Na Produção De Biocombustíveis Líquidos.* Brasília, DF: Embrapa Agroenergia.

Malaker, A., & Ahmad, S. A. I. (2013). Therapeutic potency of anticancer peptides derived from marine organisms. *International Journal of Engineering, 2*, 2305–8269.

Mali, S., Grossmann, M. V. E., & Yamashita, F. (2010). Filmes de amido: Produção, propriedades e potencial de utilização. *Semina: Ciências Agrárias, 31*(1), 137–155.

Marsic, D., Flaman, J. M., & Ng, J. D. (2008). New DNA polymerase from the hyperthermophilic marine archaeon *Thermococcus thioreducens*. *Extremophiles, 12*, 775–788.

Mayer, A. M., Glaser, K. B., Cuevas, C., Jacobs, R. S., Kem, W., Little, R. D., & Shuster, D. E. (2010). The odyssey of marine pharmaceuticals: A current pipeline perspective. *Trends in Pharmacological Sciences, 31*(6), 255–265.

Miao, X., & Wu, Q. (2004). High yield bio-oil production from fast pyrolysis by metabolic controlling of *Chlorella protothecoides*. *Journal of Biotechnology, 110*(1), 85–93.

Mitchell, S. A., & Ahmad, M. H. (2003). Establishment of *ex vitro* and *in vitro* germplasm collections of important trees and medicinal plants of Jamaica. *Jamaica Journal of Science and Technology, 14*, 9–16.

Mitsuhashi, S. (2014). Current topics in the biotechnological production of essential amino acids, functional amino acids, and dipeptides. *Current Opinion in Biotechnology, 26*, 38–44.

Morais, L. K., Cursi, D., dos Santos, J. M., Sampaio, M., Camara, T., Silva, A. P. D. A., & Gazaffi, R. (2015). *Melhoramento Genético da Cana-De-Açúcar.* https://www.infoteca.cnptia.embrapa.br/infoteca/bitstream/doc/1042764/1/Doc200.pdf (accessed on 18 March 2025).

Mota, C. J., & Monteiro, R. S. (2013). Química e sustentabilidade: Novas fronteiras em biocombustíveis. *Química Nova, 36*, 1483–1490.

Murray, P. M., Moane, S., Collins, C., Beletskaya, T., Thomas, O. P., Duarte, A. W., & Walsh, D. J. (2013). Sustainable production of biologically active molecules of marine-based origin. *New Biotechnology, 30*(6), 839–850.

Mutanda, T., Ramesh, D., Karthikeyan, S., Kumari, S., Anandraj, A., & Bux, F. (2011). Bioprospecting for hyper-lipid producing microalgal strains for sustainable biofuel production. *Bioresource Technology, 102*(1), 57–70.

Myshak, H. (2018). Definition of the term biotechnology. *Cogito: Multidisciplinary Research Journal, 10*, 142.

National Cancer Institute. (2009). *BRCA1 and BRCA2: Cancer Risk and Genetic Testing.*

Patel, M. (2006). Medium and long-term opportunities and risks of the biotechnological production of bulk chemicals from renewable resources—The potential of white biotechnology: The BREW Project. *Innovation, 6*(5), 452.

Pereira, C. G., Colla, J. N., Carminatti, L. P., Zilli, M. B., & Koenig, F. (2005). Estudo de novos polímeros biodegradáveis para a aplicação no segmento de utilidades domésticas. *CEP, 95088*, 060.

Philippe, R. N., De Mey, M., Anderson, J., & Ajikumar, P. K. (2014). Biotechnological production of natural zero-calorie sweeteners. *Current Opinion in Biotechnology, 26*, 155–161.

Phillipson, J. D. (2007). Phytochemistry and pharmacognosy. *Phytochemistry, 68*, 2960–2972.

Pinto, G., Araújo, C., Santos, C., & Neves, L. (2013). Plant regeneration by somatic embryogenesis in *Eucalyptus* spp.: Current status and future perspectives. *Southern Forests: A Journal of Forest Science, 75*(2), 59–69.

Pistelli, L., Giovannini, A., Ruffoni, B., Bertoli, A., & Pistelli, L. (2010). Hairy root cultures for secondary metabolites production. In *Bio-Farms for Nutraceuticals: Functional Food and Safety Control by Biosensors* (pp. 167–184).

Polin, B. S. (2012). *Insulin: Its History and Future.* Retrieved from: http://www.diabeticlifestyle.com/blood-glucose/insulin-its-history-future (accessed on 18 March 2025).

Ponstein, A. S., Verwoerd, T. C., & Pen, J. (1996). Production of enzymes for industrial use. In G. B. Collins & R. J. Shepherd (Eds.), *Engineering Plants for Commercial Products and Applications* (Vol. *792*, pp. 91–98). New York: New York Academy of Sciences.

Pradella, C. (2015). Biopolímeros e intermediários químicos. *Centro de Gestão e Estudos Estratégicos, 120*, 1–119.

Pulz, O., & Gross, W. (2004). Valuable products from biotechnology of microalgae. *Applied Microbiology and Biotechnology, 65*, 635–648.

Qin, S., Liu, G. X., & Hu, Z. Y. (2008). The accumulation and metabolism of astaxanthin in *Scenedesmus obliquus* (Chlorophyceae). *Process Biochemistry, 43*(8), 795–802.

Raju, T. N. (1999). The Nobel chronicles. 1945: Sir Alexander Fleming (1881–1955); Sir Ernst Boris Chain (1906–79); and Baron Howard Walter Florey (1898–1968). *The Lancet, 353*(9156), 936–936.

Rasmussen, R. S., & Morrissey, M. T. (2007). Marine biotechnology for production of food ingredients. *Advances in Food and Nutrition Research, 52*, 237–292.

Romond, E. H., Perez, E. A., Bryant, J., Suman, V. J., Geyer, J. R. C. E., Davidson, N. E., & Wolmark, N. (2005). Trastuzumab plus adjuvant chemotherapy for operable HER2-positive breast cancer. *The New England Journal of Medicine, 353*(16), 1673–1684.

Rorrer, G. L., & Cheney, D. P. (2004). Bioprocess engineering of cell and tissue cultures for marine seaweeds. *Aquacultural Engineering, 32*(1), 11–41.

Sánchez, J. F., Fernández, J. M., Acién, F. G., Rueda, A., Pérez-Parra, J., & Molina, E. (2008). Influence of culture conditions on the productivity and lutein content of the new strain *Scenedesmus almeriensis*. *Process Biochemistry, 43*(4), 398–405.

Silveira, J. M. F. J. D., Borges, I. D. C., & Buainain, A. M. (2005). Biotechnology and agriculture: From science and technology to the impacts of innovation. *São Paulo in Perspective, 19*, 101–114.

Smith, K. A. (2012). Louis Pasteur, the father of immunology? *Frontiers in Immunology, 3*, 68.

Tabassum, B., Khan, A., Tariq, M., Ramzan, M., Khan, M. S. I., Shahid, N., & Aaliya, K. (2017). Bottlenecks in commercialization and future prospects of PGPR. *Applied Soil Ecology, 121*, 102–117.

Taşkın, H., Baktemur, G., Kurul, M., & Büyükalaca, S. (2013). Use of tissue culture techniques for producing virus-free plants in garlic and their identification through real-time PCR. *The Scientific World Journal, 2013*, 781282.

Thakur, N. L., Thakur, A. N., & Müller, W. E. G. (2005). Marine natural products in drug discovery. *Natural Product Radiance, 4*(6), 471–477.

The Tomato Genome Consortium. (2012). The tomato genome sequence provides insights into fleshy fruit evolution. *Nature, 485*, 635–641.

Tindamanyire, J. M. (2014). *Optimal Growth Conditions of Agrobacterium rhizogenes for Use in the Transformation of Banana (Musa spp.)* (Doctoral dissertation, Makerere University).

Tripathi, U., Sarada, R., Rao, S. R., & Ravishankar, G. A. (1999). Production of astaxanthin in *Haematococcus pluvialis* cultured in various media. *Bioresource Technology, 68*(2), 197–199.

Verma, A. S., Agrahari, S., Rastogi, S., & Singh, A. (2011). Biotechnology in the realm of history. *Journal of Pharmacy & Bioallied Sciences, 3*(3), 321.

Villadsen, J. (2007). Innovative technology to meet the demands of the white biotechnology revolution of chemical production. *Chemical Engineering Science, 62*(24), 6957–6968.

Walsh, G. (2005). Therapeutic insulins and their large-scale manufacture. *Applied Microbiology and Biotechnology, 67*, 151–159.

Xiao, N., & Jiao, N. (2011). Formation of polyhydroxyalkanoate in aerobic anoxygenic phototrophic bacteria and its relationship to carbon source and light availability. *Applied and Environmental Microbiology, 77*(21), 7445–7450.

Yu, L., Dean, K., & Li, L. (2006). Polymer blends and composites from renewable resources. *Progress in Polymer Science, 31*(6), 576–602.

Yun, D. J., Hashimoto, T., & Yamada, Y. (1992). Metabolic engineering of medicinal plants: Transgenic *Atropa belladonna* with an improved alkaloid composition. *Proceedings of the National Academy of Sciences, 89*(24), 11799–11803.

Zhang, C. (2012). Hybridoma technology for the generation of monoclonal antibodies. *Antibody Methods and Protocols, 901*, 117–135.

CHAPTER 2

Chromosomal Theory of Inheritance

ABIDA PARVEEN,* SHAGUFTA PERVEEN, and FARAH NAZ

Department of Botany, Government College University, Faisalabad, Pakistan

*Corresponding author

ABSTRACT

This chapter explores the Chromosomal Theory of Inheritance and its role in explaining how traits are passed from parents to offspring. It begins with the early discovery of DNA and the realization that chromosomes carry genetic information. The chapter reviews key experiments that confirmed DNA as the genetic material and introduces the iconic Watson-Crick double-helix model. It explains the process of DNA replication and how it supports the transmission of genetic traits. Mendel's foundational work on heredity is connected to modern discoveries in genetics. The evolution and classification of sex chromosomes are discussed, along with the reasons why some DNA regions do not undergo recombination during reproduction. The role of mobile genetic elements is also examined. Additionally, the chapter highlights recent advances in gene analysis and genome technology. It concludes by addressing challenges within the model and how they have been resolved through scientific research.

2.1 INTRODUCTION

The evolution of genomics' background can be assembled in an array of methods. One, which is attempted here, is to demonstrate the evolution and moderate advancement in understanding of the origins of the genetic

Chromosomal and Genomic Structures and Functions, Volume 1: Chromosomes and Cell Division.
Tariq Ahmad Bhat and Jameel M. Al-Khayri (Eds)

DOI: 10.1201/9781779640253-2

material. This is a fascinating description. This approach is particularly relevant because it symbolizes the 60th celebration of deoxyribonucleic acid (DNA) molecular structure discovery, the material that makes up genes, written by British scientist Francis H.C. Crick and American scientist James D. Watson (Watson and Crick, 1953).

The well-known double-helix model, which has come to represent biology and even function as a kind of contemporary cultural emblem, was first observed by Watson and Crick in Cambridge, England, on February 28, 1953, after a year and a half of vigorous work; at the time, Crick was 36 and Watson was only 24. It is rare for any other scientific discovery to be this accurately dated, and especially not for one as significant as this one. Since then, multiple physicochemical and electron microscopy research investigations have confirmed the accuracy of the Watson and Crick model, which remains unaffected.

James D. Watson's important publication, The Double Helix (Watson, 1968), offers a fascinating account of the occurrences leading up to as well as following the discovery. The model was initially presented in a very brief article (around 800 words and one figure) on April 25, 1953. Horace F. Judson, an American historian of molecular biology, suggests that the writers' fierce competition triggered the adoption of this simplified form (Judson, 1996, p. 167). Watson, at least, considered that it came down to race. The competitors were the British biophysicist and X-ray crystallographer Rosalind E. Franklin and the United States scientist Linus C. Pauling (Watson, 1968).

Watson and Crick, however, provided a more detailed explanation of DNA structure and the model's genetic implications shortly after the model was introduced (Watson and Crick, 1953). The following year's Quantitative Biological Sciences Cold Spring Harbor Conferences offered an even more in-depth examination of the subject matter (Watson, 1954).

The Watson-Crick model is innovative and has a good explanatory capacity. It takes into account all four of the fundamental characteristics necessary for genetic information. The concept first emphasizes how genetic material replicates, which is, in turn, a foundation for reproduction and a vital component of life. Second, the model illustrates how the duplication process maintains the uniqueness of genetic material or the gene's quality. Thirdly, DNA is an informational macromolecule, which helps to explain the information contained in genetic material according to the model. Fourthly, the concept demonstrates how genetic material can alter, or how genes can mutate. We shall discuss these four essential and vital characteristics of the genetic material in more detail later.

The X-ray crystallographic work of British scientists Maurice H. F. Wilkins, his colleagues, and his countrymen Raymond G. Gosling and Rosalind E. Franklin was published in the article in Nature in the aftermath of the initial study by Franklin and Gosling (1953); Watson and Crick (1953); Wilkins (1956). The DNA model proposed by Watson and Crick (1953) was validated by these studies.

Multiple studies in electron microscopy and biological physics have now been presented, reinforcing the hypothesis.

2.2 DEOXYRIBONUCLEIC ACID (DNA) RESEARCH

The identification of DNA as a chemical entity in human leukocytes was made as early as 1869 by the Swiss doctor and scientist Friedrich Miescher. The discovery of DNA is especially important because it is associated with the finding of the principles of generational transmission by Gregor Mendel, an Augustinian priest and German-Silesian physicist who is considered the father of the modern discipline of genetics (Mendel, 1865). Miescher was fascinated by the chemical composition of the cell nucleus and conducted research in Professor Felix Hoppe-Seyler's laboratory at the University of Tübingen in Germany. Surgical bandages from the nearby hospital were used to collect pus for the research material. Leukocytes in significant quantities, with rigorously purified nuclei, are found in pus. Miescher isolated an entirely distinct organic substance from them, which he called nuclein (Miescher, 1871). This name reflected the entity's place of origin. We now understand that the material was DNA.

For example, nuclein's extraordinarily high concentration of phosphorus set it apart from all other organic molecules extracted from cells. This fascinated notice and incredulity at the same time. The most significant figure in organic chemistry at the time, Felix Hoppe-Seyler, Miescher's professor, was so uncertain that before permitting Miescher to publish the results, he wanted to conduct the experiments himself (Dahm, 2008). This caused a two-year delay in the discovery's release. Several researchers had already begun to address the challenges presented by Miescher's work during his lifetime. There have been modifications made to nucleic acid purification techniques. Miescher's student Richard Altmann, a German pathologist and histologist, assumed he had discovered a novel material that he termed nucleic acid since it exhibited an acidic nature in chemical reactions. He was not aware, meanwhile, that it was the very same material that Miescher had given the name nuclein (Dahm, 2008).

A number of other researchers were able to demonstrate a few years later that nucleic acid was an integral part of chromosomes; Edward Zacharias, a botanist, was the first to do so in 1884 (Mirsky, 1968). The discovery that nucleic acid comprised four independent bases was made by German biochemical researchers Albert Neumann and Albrecht Kossel in 1893. Furthermore, Kossel observed a correlation between chromatin and nuclein, which, combined with other proteins like histones that he had observed earlier, makes up chromosomes (Olby, 1994; Portugal and Cohen, 1977).

Nucleic acids have an important function in the synthesis of newly generated cytoplasm during development and substitution, as Kossel concluded in his research (Kossel, 1913).

Despite these important achievements, nucleic acid research was consistently less popular until the 1930s, during which time its value was primarily lost.

2.3 CHROMOSOME THEORY OF INHERITANCE

The fact that DNA is a component of chromosomes was understood as early as the 1900s, as was previously indicated. Therefore, it is helpful to start this part with a brief description of the background to the hypothesis that the biological foundation for heredity is the chromosome.

The chromosomal theory of inheritance was first proposed in 1904 by American geneticist and doctor Walter S. Sutton and German scientist Theodor H. Boveri, which holds that the chromosomes in the nucleus of the cell contain genes. This occurred not long after Mendel's inheritance laws were purportedly revealed (Boveri, 1904; Sutton, 1903).

Evidence was discovered by Boveri showing that the sea urchin *Paracentrotus lividus* has particular chromosomes with unique characteristics (Boveri, 1902). He was also able to demonstrate that a specific chromosomal combination was more significant for normal development than a specific number. He proved that chromosome integrity is a fundamental feature of chromosomes through his research on the egg divisions of the *Ascaris megalocephala* parasitic nematode worm (Boveri, 1903). Based on his studies of the sperm generation of *Brachystola magna*, he illustrated the similarities between Mendel's hypothesized separation of character variations at gamete production and the segregation of complementary chromosomes at meiotic divisions (Sutton, 1903). He further noted that the particular inheritance of unique character distinctions noted by Mendel is paralleled by the distinct orientation of every bivalent, or homologous, pair of chromosomes during the initial meiotic metaphase. Stated a certain way, he illustrated Mendel's laws of inheritance, discrimination, and independent assortment by utilizing the behavior of chromosomes during meiosis.

Thomas Hunt Morgan, an American embryologist and geneticist, and his colleagues utilized *Drosophila melanogaster* as their experimental model throughout the 1910s to demonstrate that the chromosomal theory of inheritance was reliable (Morgan, 1919, 1926; Morgan et al., 1923). Morgan first proposed that the associated trait originated on a chromosome called the X, which determines sex, in order to explain the sex dependency of the white color of eye characteristic in *Drosophila*, which he had discovered (Morgan, 1910). The chromosome hypothesis of inheritance was supported by the initial observations of the chromosomal basis of sex determination made by McClung (1902), an American naturalist and geneticist.

Following this, Morgan's student Alfred H. Sturtevant established the groundwork for gene mapping by demonstrating that the linear linkage map of genes resembled the chromosome's linear structure, supporting the chromosomal theory of inheritance (Sturtevant, 1913). Other students of Morgan, Calvin B. Bridges, noted that the nondisjunction, or lack of segregation in gamete formation, of sex-related genes in *Drosophila melanogaster* was succeeded by a compatible nondisjunction of the X chromosomes during meiosis, providing the first evidence of the theory at some time thereafter (Bridges, 1914, 1916).

More direct evidence supporting the chromosome hypothesis of inheritance was obtained in 1929 by Hermann J. Muller, an American geneticist and former student of Morgan. Muller collaborated in the laboratory of American geneticist Theophilus S. Painter and Ukrainian-American evolutionary biologist Theodosius G. Dobzhansky at Columbia University in New York. They found that following X-ray irradiation, structural alterations in *D. melanogaster* linkage groups were linked to comparable modifications in chromosomes (Dobzhansky, 1930; Muller and Painter, 1929). The arrangement of the genes on the linkage map correlates to their physical order on the chromosome, as their work ultimately demonstrated. This theory received further evidence when it was shown that actual swapping of chromosomal segments with cytological markings accompanied genetic recombination between linked markers in *D. melanogaster* and maize (Stern, 1931).

The identification of enormous chromosomes in the salivary gland cell nuclei of *Drosophila* and other dipteran larvae drastically improved the analytical capacity of cytogenetics (Painter, 1933, 1934). He used these chromosomes for the first time to create a thorough cytological diagram of the X-chromosomal region in *D. melanogaster*. He demonstrated how the symmetrical arrangement of gene blocks in the linkage map of this chromosome resembled the sequence of salivary chromosomal regions.

2.4 PROOF OF THE DEOXYRIBONUCLEIC ACID (DNA) THEORY

Mutation research utilizing *Sphaerocarpus donnellii* liverwort, specific microbiological fungus, and corn as testing species provided the initial clues indicating that the chromosomal nucleic acid part, as compared to the protein portion, would provide the genetic material (Srb and Horowitz, 1944). The UV wavelength that, upon radiation, triggered the highest frequency of alterations was found to match the largest amount that DNA could absorb. However, none of these parties was prepared to conclude that DNA was the genetic substance based only on this finding. The Americans found conclusive evidence to support the theory that genetic traits are carried by DNA (Ot, 1944). Numerous additional traits were later demonstrated to be passed down through DNA in a variety of bacterial species, including pneumococcus. Since only proteins could be expected to possess the precision required for DNA at the time, it was generally accepted that proteins were a contaminant in their preparation. Furthermore, the tetranucleotide theory, which explains DNA structure, was initially put forward by Hermann Steudel in 1906 and further examined by Lithuanian-American biologist Phoebus Levene in 1931 (Deichmann, 2004).

However, in the 1930s, genuine understanding of the structure of DNA also grew after the proposal of English molecular biologist and physicist William T. Astbury that DNA was a long, uniform molecule with loops.

Based on the substance utilized for X-ray crystallography, he made this determination (Astbury and Bell, 1938). The work of American geneticists Alfred D. Hershey and Martha Chase, which demonstrated that DNA was the cause of bacteriophage proliferation, was more persuasive to the scientific community as a whole (Hershey and Chase, 2017), demonstrating that genetic traits might be inherited only by DNA, as there doesn't seem to be any correlation between the process of genetic transmission and protein. These findings revealed that the protein component of the chromosome may have a different purpose in higher forms of life and that DNA alone may represent the molecular foundation of inheritance in those forms as well. The French scientist André F. Boivin had proposed this in 1948 based on the base makeup of DNA and the amount of DNA in nuclei (Boivin et al., 1948). It is now well accepted that DNA is the genetic substance shared by all living things on Earth, with the notable exception of RNA viruses.

Erwin Chargaff, an Austrian biochemist, was part of a small group of scientists who acknowledged the significance of the research conducted by Oswald Avery and his colleagues, accepted its conclusions as infallible, and proceeded accordingly. In the late 1940s, Chargaff and his associates demonstrated the specificity of the DNA structure while also showing that the tetranucleotide theory needed to be incorrect (Chargaff, 1950).

2.5 GENETIC IMPLICATIONS OF THE MODEL

The qualities required for genetic material—auto-replication, specificity, and information content—were satisfied by their model of the DNA structure. Furthermore, the model takes into consideration the possibility of genetic material changing, or the capacity of genes to shift. Chargaff's theory of base-pairing in DNA's structure provides the basis for the auto-replication of DNA (Chargaff, 1950). Thus, adenine and cytosine are inherently antagonistic to thymine and guanine, respectively. Because of its linear shape, DNA possesses specificity and information content; DNA's nucleotide sequence contains genetic information. An essential prerequisite for biological evolution is the capacity of genes to change, which is based on the intrinsic characteristic of the hypothesis that DNA nucleotides can occasionally be altered. Nevertheless, during development, animals may rely on the information stored in DNA since its structure is sufficiently durable. These features of DNA are prerequisites for the DNA necessary for life's development. Even that molecule has to possess these qualities if life exists elsewhere in the cosmos and its foundation is not DNA, but rather another large molecule.

2.6 CHALLENGES WITHIN THE MODEL AND THEIR RESOLUTIONS

Despite its beauty, the double-helix structure presented several significant physico-chemical challenges. As Watson and Crick have already indicated, they determined that synthesis then spread out in both directions, creating a "replication bubble" that finally merged with its nearby bubbles, and that replications started at various locations along the chromosome, unwinding (Watson and Crick, 1953). This finding solved the unwinding problem and hypothesized that the two strands of DNA are side by side rather than coiled around one another when it was first proposed at the end of the 1970s (Huberman and Riggs, 1968). The double-helix model was not the only model that suited the X-ray crystallographic data even better than some of these others. In addition, these side-by-side versions packed up easier than the double-helix variant since they were tape-like rather than rope-like and required no trouble unwinding. As a result, these models represented a notable contrast and established a substantial body of evidence supporting the double-helix model at the time, and they were able to refute the hypotheses that were based on how helical DNA molecules moved in the electrophoresis solution. Each daughter DNA molecule in partial replication has a double helix made up of one freshly synthesized strand and one strand from the parent DNA molecule. Alternatively, the parent molecule is conserved in conservative replication, and a single daughter double helix made up of two freshly made strands is created. Daughter molecules are made up of strands that contain portions of both the parental and freshly generated DNA during dispersive replication (Crick, 1979).

2.7 MECHANISM OF DEOXYRIBONUCLEIC ACID (DNA) REPLICATION AND DEVELOPMENT OF GENETIC

The procedures for DNA replication are distinct in the assumptions they develop, affecting the allocation among offspring molecules of atoms that originate from parent molecules, they observed. To ascertain the parental tag dispersion throughout offspring DNA molecules, one only required to label an organism's DNA and facilitate its replication in another environment that does not require marking. The bacteria *Escherichia coli* were essentially given nutrients with a heavy isotope of nitrogen (15N) over several generations (Matthew Meselson and Franklin W. Stahl, 1958). With only light DNA progenitors present in media sources, the microbes were allowed to continuously reproduce once, twice, and three times when 100% of the nitrogen in their DNA belonged to the heavy isotope. DNA samples from certain replication cycles were conducted in an ultracentrifuge using a cesium chloride gradient. A hybrid molecule created by one replication generation has a band midway between the heavy and light DNA (Falk, 2009). The chromosome-forming molecules, or DNA, do not, however, automatically exhibit this replication pattern since, at the time, there was no proof to support the theory that each chromatid was made up of a single complementary DNA strand. In contrast to this, the trial (Matthew Meselson and F. W. Stahl, 1958).

According to a theory put forth by Crick, the process of protein production is biphasic. The first stage, known as genetic transcription, involves copying the genetic data found in genes to RNA within the nucleus. Following that, this is transferred into the cytoplasm, where it is transformed into an amino acid sequence during genetic translation, which is the second step of protein synthesis. Now, the intervening RNA is responsible for carrying this. Cells' genetic information can only go from DNA to RNA and then to proteins; it is unable to move from proteins to nucleic acids. The basic principle of Molecular Biology states this, which Crick initially proposed in the same publication. Proteins cannot exchange information with one another. In 1970, Cecil B. Davis introduced an improved version of this fundamental principle (Crick, 1970). Though the concepts had not yet solidified, RNA was already being seriously explored in the late 1950s as the mediator between DNA and proteins. An initial hypothesis regarding the transfer of information from genes to the cytoplasm postulated that a gene would result in the production of a single, specialized kind of protein (Brenner et al., 1961).

According to their investigations on the control of the lactose operon in *E. coli*, the idea that messenger RNA (mRNA) plays a crucial role in protein synthesis was unequivocally validated very soon after by a number of investigations carried out by various researchers in the early 1960s (Portin, 1993). A significant contribution to the resolution of the genetic code puzzle came from pure genetic mutation research carried out by Francis Crick and colleagues once again (Crick et al., 1961). This research revealed that the genetic code, which determines the correlation between protein and gene sequences, is a triplet code without commas, where the coding words occur one after the other rather than simultaneously. This is a first for the code. "Comma-less" refers to the absence of nucleotides between codons, which function as punctuation marks, while "triple code" refers to the structure of code words, or codons, in DNA. Deciphering the genetic code biochemically *in vitro* was achieved shortly thereafter, between 1961 and 1965. Put otherwise, the DNA code triplets that matched were found to correspond to each amino acid that is a part of proteins' fundamental structure. Regarding this project, Marshall W. Nirenberg, H. Gobind Khorana, J. Heinrich Matthaei, and Severo Ochoa all achieved notoriety while working in the U.S. at the time. The greatest publications have detailed the sections of this outstanding accomplishment in molecular biology (Portin, 2014).

They looked at mutations in the *E. coli* bacteria's tryptophan synthetase gene before reaching this result. The sequence of mutations on the gene's fine structure map was compared to the sequence of changes in the enzyme protein's primary structure triggered by these mutations. These orderings were found to be consistent with one another (Yanofsky, 1963). Additionally, it was shown that assuming single-nucleotide changes in the related gene and the validity of the DNA code *in vivo* may account for the influence of the mutations on the sequence of amino acids. The same theory states that the physiological impacts of intragenic recombination on the sequence of amino acids may be explained by recombination among two nucleotides and the integrity of the genome's code *in vivo* (Yanofsky et al., 1966).

Furthermore, Heinz-Günter and Brigitte Wittmann Liebebold, two German biochemists, conducted mutation research utilizing TMV, or tobacco mosaic virus, as the genetic coding was validated *in vivo* by the experimental organism. Nitrous acid was used to produce mutagenic mutations in TMV, an RNA virus with completely distinct mutagenic properties. Specifically, it converts adenine to hypoxanthine and cytosine to uracil, which then pairs with guanine to form RNA mutations (uracil being one of the bases of RNA, not thymine). The impact of 24 mutations on the virus's coat protein's amino acid sequence was examined. Based on the single-nucleotide mutations indicated and the assumption that the genetic code was intact in life, 23 of these occurrences might be explained.

It was determined that the lone remaining example was a spontaneous mutation (Wittmann and Wittmann-Liebold, 1966). It was discovered that the genetic code is universal. This indicates that almost every living thing on the planet, including viruses, bacteria, fungi, plants, and animals, uses the same code. This is one of the most compelling arguments in favor of the theory of evolution. There are hardly any exceptions to this rule. For instance, a few code triplets have distinct meanings in the nuclear and mitochondrial genomes.

2.8 BIOCHEMICAL GENE ANALYSIS AND GENOME TECHNOLOGY

The genetic information from both a virus and a bacterium was simultaneously transferred to the genome of a distinct viral species. Employing the technique developed by Berg and colleagues, distinct recombinant DNA molecules, including DNA from several sources, may be purposefully created. For example, human DNA may be transferred into a bacterial cell and allowed to amplify by inserting a specific DNA sequence into a molecule that is recombinant, utilizing certain gene vectors. Alternatively, amplification can be achieved by the use of the polymerase chain reaction (PCR), which was developed in 1983 by American biochemist Kary Mullis (Bartlett and Stirling, 2003). A sizable portion of the relevant DNA, adequate for a variety of chemical and biological investigations and uses, may be produced using either method. For example, using restriction mapping, which was developed by British genetic scientist Edwin Southern (Southern, 1975).

However, the majority of these later, much faster, and less expensive DNA sequencing techniques share the same fundamental idea as these earlier techniques: specific points in DNA are broken, and the fragments are then separated according to their sizes using electrophoresis or a similar technique. We can see the sequencing process unfold in real time thanks to the most recent sequencing techniques. The most recent of these techniques make use of nanotechnology. The most precise physical gene mapping that is currently possible is achieved by genome and gene sequencing. Furthermore, the genomes of prokaryotes such as bacteria and the whole genomes of vast numbers of species of eukaryotes have been thoroughly examined in recent years. This has resulted in a radical shift in the direction of biology research across the board, including biomedical research. The above-described victorious march of molecular genetics was made possible by the discovery of DNA's structure 60 years ago. These discoveries and innovations have earned the composers of these works several Nobel Awards. The release of the whole human genome sequence, initially in draft form in February 2001 (Subramanian et al., 2001).

An important illustration of the significance of the Human Genome Project and related initiatives pertaining to our closest relatives, both extant and extinct, is the profound comprehension that resulted about the connections between current populations of humans and their ancestors. The genome sequence of chimpanzees and a comparative analysis with the genome of humans were made public (Chimpanzee Sequencing and Analysis Consortium, 2005). A young hominin's distal manual phalanx was discovered during an excavation at Denisova Cave in the southern Siberian Altai Mountains. This discovery led to the identification of a new extinct human species that was later named Denisova man. The ordering of the genome (Reich et al., 2010) is significant. From the perspective of human natural history, I think the following are the most significant discoveries in this field: Firstly, most of our inherited traits are shared with other primate species, around 60% with all animals, about 80% with vertebrate organisms in general, and about 20% with all things that exist (BAC Resource Consortium et al., 2001). The enormous shift in our knowledge of the structure and function of the genome that resulted from the massive sequencing programs is another extremely important effect (Portin, 2009).

2.9 MENDELS WORK ON HEREDITY

The Mechanism of Mendelian Heredity, published in 1915 by T. H. Morgan and coworkers at Columbia University, was the result of multiple substantial periods of fruit fly breeding investigation using *Drosophila*. They put forth a hypothesis that made an association between the behavior throughout meiotic splits of chromatin and the segregation of genetic components in inheritance with Mendelian heredity (Morgan, 1915). Then, they discovered that sets of characteristics, the quantity of which was comparable to the number of haploid chromosomes in their test organism, could be "linked" and inherited.

Cytological evidence was used to establish the validity of chromosomal crossing-over, and breeding data was utilized to predict the probability of these incidents. They suggested using chromosome tracking to predict and identify the precise locations of certain genetic components. This might be achieved by carefully selecting fly breeds to crossbreed. Many biologists at the time argued that the evidence proving the model's mechanism was "overpowering," as each component's efficacy added up to support it (Brush, 2002). They added that a wide range of living things could benefit from their chromosome hypothesis of inheritance.

Many of the premises of Morgan's model depended upon a small number of completed observations in one organism, which is why British biologist William Bateson, who originally established the new discipline of "genetics," vehemently resisted the model's implementation. Though it was unproven, he predicted the work to be instructive (Cock, 1983). A number of researchers have endeavored to comprehend Bateson's expressive and persistent avoidance of the chromosomal theory. A cynic noted that Bateson initially accepted the idea two weeks prior to his death in 1926 (Hutchinson, 1979).

William Coleman thought Bateson's antagonistic nature was a result of his character. Morgan's argument that the entire explanation of life might be found in terms of chemical and physical laws did not sit well with Bateson, who was politically moderate, anti-materialist, intuitive, and aesthetic (Coleman, 1971). Bateson was an experienced scientist who struggled to grasp novel concepts in genetics, according to Alan Cock. Additionally, he rejected Morgan's depiction as "unfit and disorganized" and showed little curiosity in using cytology to determine heredity.

Hereditary components, described by Lindley Darden and Alan Rushton, are not material fragments; instead, they are mechanisms, conduits, and whirlpools that govern the growth, division, and reproduction of biological life. However, Bateson wasn't able to translate these ideas into a testable and proven genetic theory (Rushton, 2014).

Interest in using cytological facts to draw conclusions about Mendelian segregation has increased after it was discovered that a mechanism providing the necessary machinery for segregation could occur during the development of germ cells. This discovery has sparked interest in using cytological facts to draw conclusions about Mendelian segregation (Morgan, 1910). The eyes of *Drosophila* flies were typically crimson. An analysis of the correlation between this trait and the insect's sex was inspired by the unexpected emergence of a male fly with white eyes. Bridges went on to say that there is no question that the X chromosome is the source of the sex-linked genes because of the full similarities between the unique ways in which chromosomes behave and the ways in which sex and sex-linked genes behave in this instance (Bridges, 1914). This work had a significant impact. Years later, Morgan noted that it was becoming more and more absurd to reject the chromosomal idea once Bridges' research was made public (Mayr et al., 2018).

Subsequently, Morgan and his associates revealed that occasionally, linked characters did not separate collectively in the expected test-cross ratios. They proposed that occasionally, chromosomes broke apart and re-annealed, a process called "crossing over." According to Sturtevant, the rates of crossing over should be proportionate to the relative distance between two genetic variables. Using this method, he created a linear chromosomal map in the fly for six sex-linked characteristics (Sturtevant, 1913). Mendel believed that every genetic element segregated on its own, but Morgan has since offered an expansion of Mendelism that does not in any way refute Mendel's ideas of segregation. Factors that are located on distinct chromosomes freely sort and produce Mendelian expectations; however, factors that are located on the same chromosome are considered to be connected and cause variations in Mendelian ratios. According to my proposal, the variations might be interpreted as the division of the relevant variables (Morgan, 1913).

Following his presentation to Bateson of his intriguing research on the chromosomes and personalities of *Drosophila*, Morgan volunteered to ship several mutant flies to England to encourage breeders to look at them. However, at this point, Bateson showed little curiosity in the relationship between heredity and cells, and he didn't forward this study opportunity to his Cambridge colleagues (Darlington, 1981).

Doncaster realized that the chromosome might be thought of as a sequence of characters, with paired characters separated together, as did Morgan. He did, however, issue a warning: while there is a lot of indirect evidence supporting the Morgan theory, which holds that each and every Mendelian feature is inherited, it cannot be taken as unquestionable. Such a physical paradigm was strongly supported by the integration of sex-specific traits and chromosomes. The recognition that male-only sex-limited transmission occurs exclusively in groups (mammals and Diptera) when the male chromosomes are unattached or sparsely paired, out of the two categories (birds and *Lepidoptera*), suggests that the inheritance of sex-limited traits and a chromosome are intimately related, as two instances in the former may now be characterized as having an unusual component in the female (Doncaster, 1914).

Though we remain doubtful about the validity of the cross-over theory and particularly the reasoning behind assigning factors, it is challenging to map to sequential places in the chromosomes. This argues against the notion that the qualities that gender-neutral scripts possess a unique relationship with the sex chromosomes (Rushton, 2014). I cannot help but feel that the research confirms the chromosomal causal link almost exactly, if not exactly, and components, but I agree that we don't really know what it means for a character to be "support" or "transported" by a chromosome. The important issue, in my opinion, is the tangible guarantee that genetic factors affect the inherited qualities in some way while the genes or components are arranged in an ordered set across a line, regardless of how crossing-over is impacted. The Morgan model for Doncaster was validated by other papers on *Drosophila*. Late in 1916, he wrote to Bateson, "the primary information appears to indicate that the components are organized into groups, but it also suggests that people are really more aware of where each factor is than previous research would have us believe." Doesn't it appear that the linear layout with constant lengths is what he was asking Bateson (Rushton, 2022)? The assumption that the *Drosophila* chromosomes truly contain linked character groups is thus highly likely due to the strong correlation between the groupings of characters.

In his book Mendel's Principles of Heredity, Bateson additionally addressed the topic of related trait inheritance, noting that it was unclear if unit character segregation occurred within a single meiotic cell division (CD). A design comprising several dividing cells that produced embryos in a variety of spatial patterns may have been worked out by numerous germ cells. However, he conceded that there were insufficient oogonia in the peas and maize plants to maintain a "replica in such a small number" (Bateson and Mendel, 2013). The tight system of symmetry that holds cells together is only present during the initial stages of development. Undoubtedly, the biggest challenge is that I am unable to observe how the differentiated cells are dispersed throughout the plant to their respective locations (Letter to Erwin Baur). Elements would have eight quadrants for three components, and so forth, and four quadrants for CD. Indeed, the model was intricate. Although this option is remote, the authors argued that even unlikely solutions should be taken into account for a challenge of this magnitude of complexity. Nevertheless, the scientists acknowledged that "we find it inconceivable that a plant's characteristic form would not alter" because replication most likely took place in each separate creature at different phases of embryonic development (Rushton, 2014).

The notion of several large-scale CDs felt like more of a byproduct that "could not occur during the moment, match very happily" with findings from experimental cytology, even if the model seemed to suit certain breeding data. Furthermore, Sturtevant said that the reduplication model failed to provide a straightforward technical justification for every visible finding and was unable to forecast the outcome of specific crossings. The trillion-plus figure, which is obviously "express inanity," is what he arrived at after calculating the quantity of cells required for duplication to split a sizable number of unit individuals. The speaker came to the conclusion that "no foundation is provided by actual (cytological) inspection for this manner," forcing them to imagine an incredibly complicated series of multiple differentiating splits of cells that occur in succession (Sturtevant, 1914). Fruit flies were used to demonstrate the connection among one chromosome, gender selection, and sex-related genes. They then included genetic mapping and crossing-over in their model.

The outcomes of a single fly crossing were precisely predicted by the model. Bateson agreed that chromosomes play a part in the sex-determination scenario but declined to work with Morgan on *Drosophila* breeding (Bateson, 1919).

2.10 EVOLUTION OF HETEROMORPHIC SEX CHROMOSOMES

Histone modifications, which are post-translational changes to a histone "tail," and DNA methylation, which is a 5-carbon cytosine nucleotide (5-methylcytosine) formed by adding a methyl group (CH_3), are examples of epigenetic processes that are recognized to have an impact on the makeup of chromatin and the activity of gene transcription. The relationships between sex chromosomal divergence, genetic changes, and crossover inhibition remain unknown, though (Gorelick, 2003). A convincing idea combines Muller's ratchet with epigenetic modification. Prior to one of both identical chromosomes resulting in the heteromorphic sex chromosome (Y or W), putative different methylation patterns would be seen in the sex-determining region of the same chromosome. Differential methylation may lead to chromatin changes that are reduced by recombination, as well as the expression and development of traits peculiar to one sex in the heterogametic sex. Second, methylated cytosines deaminate to form thymine's more quickly than unmethylated cytosines due to their hypermutability. This spatially differential methylation would therefore quicken the methylation and increase Muller's ratchet by increasing the frequency of variation in the dispersion of the sexual chromosomes. TPG sites replace CPG sites (Holliday and Grigg, 1993).

Estimating the DNA at the start of sex genomic separation, the gender-determining region's methylated status is difficult since DNA methylation may regulate gene expression in a complicated way (Jaenisch and Bird, 2003). As an illustration, transcriptional repression is linked to the hypermethylation of promoter regions. On the other hand, active transcription is linked to hypermethylation in the gene body, based on whether these elements are enhancers or repressors. Additionally, legislative elements—which are located many megabases away from the genes they regulate—can have their function controlled by the effects of DNA methylation. These elements may have opposing effects on a gene (Jaenisch and Bird, 2003). Problems might also arise from the necessity of inherited DNA methylation patterns. These methylation concentrations are maintained during fertilization and development in certain species (like zebrafish) but are eliminated during development in other organisms (like humans). Since it is not well known how the conversion and restoration of DNA methylation occur throughout infancy, it is difficult to generalize concerning the influence of remodeling the methylation of DNA upon the inheritance of gained habits of methylation. However, it's feasible that animals without the aforementioned resetting have a more likely explanation for the potential role of chromatin processes in the development of sex chromosomes. When combined, these concepts make Gorelick's hypothesis difficult to evaluate. This idea relies on greater amounts of methylation because of its impact on recombination, mutation rates, and chromatin structure. When a gender-determining genome manifests as a heterogametic sex, Gorelick's theory of a required epigenetic sequence skewed toward sex would not be satisfied by the traditional method of regulating gene expression through promoter methylation. Under these circumstances, one may anticipate that only within gene regions will hypermethylation of the sex-determining locus occur. As an alternative, Gorelick's idea would be supported by a higher methylation influence of the phase-determining gene, which is linked to lower activity in the heterogametic sex. The promoter domain in the heterogametic sex in a phase-determination process, along with the extent of the sex genes' distinction from the technique employed to detect DNA methylation, both affect the predicted DNA methylation pattern of the sex-determining gene. The homogametic sex (XX females or ZZ males) DNA is used to sequence and assemble a large number of reference genomes. As a result, if the presence of Muller's ratchet is accelerated by methylation of cytosines in the non-recombining region, then the conversion of methylated cytosines to thymines would reduce how often CpG sites are found in the non-recombining zone of heteromorphic chromosomes. This means that it would appear that the homosexual sex is more extensively methylated than the heterogametic sex in the non-recombining area among separating heteromorphic sex chromosomes. When examining whole-genome bisulfite sequencing (WGBS) data, contrary to what could initially seem to be predicted, this pattern is compatible with what should be expected, as shown earlier in a highly evolved sex chromosomal system reported when employing a bisulfite sequencing technique (Metzger and Schulte, 2018).

Some of the more intriguing mysteries relating to the evolution of sex chromosomes in non-model species and epigenetic processes are starting to be addressed thanks to significant new developments in sequencing technologies. In conjunction with information on DNA methylation or histone modifications, the use of methods that offer chromosomal architecture (such as ATAC-seq, ChIP-Seq) surrounding a locus that determines sex may offer functional evidence for an epigenetically driven process. So, it makes natural sense in many respects that the degree of sex chromosomal differentiation might vary within a species. Even if adaptive mechanisms like sexual conflict contribute to the explanation of sex chromosomal differentiation, it is still not entirely clear (Charlesworth, 1991). Periods of polymorphism result from the time it takes for these variations to settle within a species. Furthermore, it's conceivable that the intensity associated with sexual tension varies among populations that have distinct behavioral environments, which causes variance in the degree of sex chromosomal differentiation. A recombination-suppressive trait may not have had enough time to fix over a species range in immature, homomorphic sex chromosomes.

Interspecific data have often been the subject of comparative research aiming to evaluate different hypotheses of sex chromosomal development

(Pokorny and Smith, 1977). Trying to take advantage of the extraordinary variety seen across several large clades, with ongoing exchange of genes and the increasingly ancient common ancestor, it is possible that comparing populations within a species might test hypotheses of sex chromosome evolution more effectively than comparing populations between species. This is because there may be fewer additional variables to take into account. The diversity seen among taxa can offer strong evidence on the possible cause of the suppression of sex chromosomal recombination. The fitness implications of chromosomal rearrangements such as inversions are essentially neutral and are unusual events that require years to correct among members of the same species, particularly when there is no dispute over sexuality (Ironside, 2010). Shifting recombination cold spots may promote the inhibition of fusion, according to the hypothesis that may be made by evaluating whether variations in lineage variation differ between individuals in the crossover patterns. In general, this variation can be used to assist in ruling out recombination suppression candidates that appear to be clear when acknowledged and used (Reichwald et al., 2015).

2.11 SEX CHROMOSOME CLASSIFICATION

Two primary kinds of sex chromosomal systems exist for species that express sex during the diploid phase. Stevens made the initial discovery of the XX/XY system, in which males have access to the Y chromosome and are heterogametic with both X and Y chromosomes (Wilson, 1906). There is also a XX/X0 system, which is a variation of this system in which males have one fewer chromosome than females. In addition, there exists an opposite system known as female heterogamete, or ZW/ZZ, wherein females are linked to the W chromosome. Z0/ZZ refers to cases where, compared to males, females carry one fewer chromosome. It's common knowledge that XX/X0 and Z0/ZZ systems arise from the deletion of an X or Z chromosome dosage-based characteristic, while the Y or W chromosome eventually determines sex (Kuroiwa et al., 2010). With two chromosomes, U and V, every individual in these systems is heterogametic during the diploid phase. During the haploid phase, individuals have both a U and a V chromosome. Several fungi also have more intricate systems in place, wherein twin homozygous chromosomes with distinct mating types must collide for recombination to occur. Multiallelic systems are used to define genetically separate mating types (Nieuwenhuis and James, 2016).

While we discuss things we may gain from various structures, including the genital kind loci of fungus, everyone focuses our research primarily on how the dynamics of evolution affect ZW and XY systems. Apart from the many varieties, sex chromosomes can also be heteromorphic, meaning that there may be some genetic diversity; for example, across the sex chromosomes, there may be SNPs, inversions, or reductions. On the other hand, homomorphic sex chromosomes might show very little divergence between the pairs. While many assessments of heteromorphy rely on whether or not genomic traits change noticeably between males and females, there is currently no agreement over the distinction between homomorphic and heteromorphic behavior in a pair of sexes. The level of divergence and sex chromosomal age are not always related (Wright et al., 2016).

Some species, such as mosses and algae, can also experience genetically determined sex during their haploid phase of life. In these instances, the sex chromosomes are known as U and V (Wright et al., 2016).

2.12 GENETIC RECOMBINATION

A common theory of recombination suppression is that all of the included loci's recombination is stopped at once by chromosomal inversions that span the sex-determining locus and other loci (Charlesworth et al., 2005). Depending on the duration of the inversion, discrete areas with varying ages and levels of degeneration are produced—a process known as stratification. Generally speaking, strata are areas where genetic traits aggregate together into unique clusters. The highest concentration of biological separation of the gender chromosomes is seen in the earliest layer, which usually represents the original recombination suppression event. A lower degree of divergence characterizes younger evolutionary strata. As a result, investigations of the evolution of X-Y counterparts in therian mammals first identified strata, which clustered into many distinct groups throughout the chromosome in a geographical manner (Lahn and Page, 1999). But there isn't much concrete proof that inversions on sex chromosomes really stop recombination. Crossover suppression was the ancestral condition, as several recent comparative genomic investigations of fungi have persuasively shown, and inversions are a result of stopped recombination rather than its fundamental cause (Grognet et al., 2014). Since the absence of recombination results in the lack of selection to preserve order, inverted poses are inclined to happen following various forms of fusion repression. The idea that tight stratum borders caused by invasions is supported by detailed investigations in several well-studied sex cytogenetic systems, which instead point to a slower and continually changing process leading to the growth of the nonrecombining zones (Cotter et al., 2016).

The notion of strata, which identifies areas with genetic traits that cluster geographically, is still relevant, but it may be more accurate to think of the boundaries between strata as hazy rather than well-defined. Moreover, developing the progression of dispersion between the sex chromosomes varies (Almeida et al., 2020). These contradict the idea that there can be only one reversal incident, suggesting that combination inhibition may be partially initial and develops through a more gradual process. Divergence patterns between the sex chromosomes will remain different in the event of discrete and progressive recombination suppression; nevertheless, it is important to note that gradual extension utilizing the non-recombining region together with X-Y codon-limited selection might provide a false stratum indication. Genetic sequences that can replicate and insert themselves into other parts of the genome are called transposable elements (TEs). The placement of TEs in proximity to the locus that determines sex might potentially reduce rearrangement by resulting in the sex carriers' separation, despite the common belief that this occurs after recombination suppression. Suppression of regions for crossover close to TE additions could ensue from this, since host mechanisms would be triggered to quiet TEs (Almeida et al., 2020). More TEs may be inserted since there is less selection against them after recombination is established, which causes the TEs to accumulate. TEs are located at the borders between recombining and non-recombining areas in regions of repressed crossing that have been found recently in birds and mammals, suggesting a causal function for them. Xu et al. (2019) could be essential in the initial phases of recombination suppression. It is plausible that TEs might concurrently stimulate sex chromosome turnover and both initiate junction inhibition and sex-determining gene degradation, given their capacity to rearrange genomes and alter variations in transcription (Ponnikas et al., 2018).

May imply that certain recombination suppression methods are reversible. Crossover events for a large number of eukaryotes are centered in hotspots (Baker et al., 2017). After finding DNA motifs, the recombination machinery is brought in. Rapid alterations occur in these binding motifs, which are primarily eliminated in favor of (Myers et al., 2010). So, a recombination cold spot might be encouraged by any selection that prevents their reestablishment on a sex chromosome. On the other hand, it is recognized that chromatin shape is regulated by genetic methylation, histone modifications, and transcription, although there are complicated and context-dependent interactions between these two epigenetic processes (Cedar and Bergman, 2009). The correlation among changes in histones and the methylation of DNA is not well understood, but it is generally known that transcriptionally repressed chromatin is associated with hypermethylation of DNA and trimethylation of histone H3 at lysine 9 (H3K9Me3), which can decrease recombination over a significant portion of the chromosome. In sticklebacks and papayas, nonrecombining sections of the sex chromosome were linked to high levels of DNA methylation, despite the lack of clear evidence connecting epigenetics to recombination suppression (Zhang et al., 2008), indicating that a possible function for DNA methylation is in recombination suppression.

Overall, it is unclear and frequently disregarded how epigenetic changes and the development of sex chromosomes are related, but some have suggested that DNA methylation may be crucial to the generation of heteromorphic sex chromosomes (Gorelick, 2003).

Additionally, methylation variations between the sexes may permit differentiating the expression of these shared genes, as the methylation of DNA affects homomorphic sex chromosomes, which are almost identical genetically. This can collaborate with other factors that affect the pace at which sex chromosomes recombine. This could foster an atmosphere that encourages sex chromosomes to continue differentiating. The inhibition of TEs by DNA methylation alterations may initiate a chain reaction in areas located on the sex-limited Y or W genome, leading to increased mutations and reduced transcription, as was previously stated. TEs may be essential to the early phases of sex chromosome development (Zamudio et al., 2015).

KEYWORDS

- **DNA copying process**
- **DNA structure**
- **gene inheritance**
- **gene mixing (recombination)**
- **gene switches (methylation)**
- **sex chromosome types**
- **Watson and Crick model**

REFERENCES

Almeida, P., Proux-Wera, E., Churcher, A., Soler, L., Dainat, J., Pucholt, P., Nordlund, J., Martin, T., Rönnberg-Wästljung, A. C., & Nystedt, B. (2020). Genome assembly of the basket willow, Salix viminalis, reveals earliest stages of sex chromosome expansion. BMC Biology, 18(1), 1–18.

Astbury, W. T., & Bell, F. O. (1938). X-ray study of thymonucleic acid. Nature, 141(3573), 747–748.

BAC Resource Consortium, T., Cheung, V., Nowak, N., Jang, W., Kirsch, I., Zhao, S., Chen, X. N., Furey, T., Kim, U. J., & Kuo, W. L. (2001). Integration of cytogenetic landmarks into the draft sequence of the human genome. Nature, 409(6822), 953–958.

Baker, Z., Schumer, M., Haba, Y., Bashkirova, L., Holland, C., Rosenthal, G. G., & Przeworski, M. (2017). Repeated losses of PRDM9-directed recombination despite the conservation of PRDM9 across vertebrates. eLife, 6, e24133.

Bartlett, J. M., & Stirling, D. (2003). A short history of the polymerase chain reaction. PCR Protocols, 226, 3–6.

Bateson, W. (1919). Studies in variegation. I. Journal of Genetics, 8(2), 93–99.

Bateson, W., & Mendel, G. (2013). Mendel's Principles of Heredity. Courier Corporation.

Boivin, A., Vendrely, R., & Vendrely, C. (1948). L'acide désoxyribonucléique du noyau cellulaire, dépositaire des caractères héréditaires—Arguments d'ordre analytique. Presse Médicale, 7, 497–507.

Boveri, T. (1902). Über mehrpolige Mitosen als Mittel zur Analyse des Zellkerns. Verhandlungen der Physikalisch-Medizinischen Gesellschaft zu Würzburg (NF, 35), 67–90.

Boveri, T. (1903). Über die Konstitution der chromatischen Kernsubstanz. https://www.semanticscholar.org/paper/%C3%9Cber-die-Konstitution-der-chromatischen-Boveri/42f4f862f0f9131ac07311cec1ea430b7aaab935 (accessed on 18 March 2025).

Boveri, T. (1904). Ergebnisse Über Die Konstitution Der Chromatischen Substanz Des Zellkerns. Verlag von Gustav Fischer in Jena.

Brenner, S., Jacob, F., & Meselson, M. (1961). An unstable intermediate carrying information from genes to ribosomes for protein synthesis. Nature, 190, 576–581.
Bridges, C. B. (1914). Direct proof through non-disjunction that the sex-linked genes of Drosophila are borne by the X-chromosome. Science, 40(1020), 107–109.
Bridges, C. B. (1916). Non-disjunction as proof of the chromosome theory of heredity (concluded). Genetics, 1(2), 107.
Brush, S. G. (2002). How theories became knowledge: Morgan's chromosome theory of heredity in America and Britain. Journal of the History of Biology, 35(3), 471–535.
Cedar, H., & Bergman, Y. (2009). Linking DNA methylation and histone modification: Patterns and paradigms. Nature Reviews Genetics, 10(5), 295–304.
Chargaff, E. (1950). Chemical specificity of nucleic acids and mechanism of their enzymatic degradation. Experientia, 6(6), 201–209.
Charlesworth, B. (1991). The evolution of sex chromosomes. Science, 251(4997), 1030–1033.
Charlesworth, D., Charlesworth, B., & Marais, G. (2005). Steps in the evolution of heteromorphic sex chromosomes. Heredity, 95(2), 118–128.
Chimpanzee Sequencing and Analysis Consortium. (2005). Initial sequence of the chimpanzee genome and comparison with the human genome. Nature, 437(7055), 69–87.
Cock, A. G. (1983). William Bateson's rejection and eventual acceptance of chromosome theory. Annals of Science, 40(1), 19–59.
Coleman, W. (1971). Bateson and chromosomes: Conservative thought in science. Centaurus, 15(3), 228–314.
Cotter, D. J., Brotman, S. M., & Wilson Sayres, M. A. (2016). Genetic diversity on the human X chromosome does not support a strict pseudoautosomal boundary. Genetics, 203(1), 485–492.
Crick, F. (1970). Central dogma of molecular biology. Nature, 227(5258), 561–563.
Crick, F. (1979). Split genes and RNA splicing. Science, 204(4390), 264–271.
Crick, F., Barnett, L., Brenner, S., & Watts-Tobin, R. J. (1961). General nature of the genetic code for proteins. Nature, 192, 1227–1232.
Dahm, R. (2008). Discovering DNA: Friedrich Miescher and the early years of nucleic acid research. Human Genetics, 122, 565–581.
Darlington, C. D. (1981). Genetics and plant breeding, 1910–80. Philosophical Transactions of the Royal Society of London. B, Biological Sciences, 292(1062), 401–405.
Deichmann, U. (2004). Early responses to Avery et al.'s paper on DNA as hereditary material. Historical Studies in the Physical and Biological Sciences, 34(2), 207–232.
Dobzhansky, T. (1930). Translocations involving the third and the fourth chromosomes of Drosophila melanogaster. Genetics, 15(4), 347.
Doncaster, L. (1914). Memoirs: Chromosomes, heredity, and sex: A review of the present state of the evidence with regard to the material basis of hereditary transmission and sex-determination. Journal of Cell Science, 2(236), 487–521.
Falk, R. (2009). Genetic Analysis: A History of Genetic Thinking. Cambridge University Press.
Franklin, R. E., & Gosling, R. G. (1953). Molecular configuration in sodium thymonucleate. Nature, 171(4356), 740–741.
Gorelick, R. (2003). Evolution of dioecy and sex chromosomes via methylation driving Muller's ratchet. Biological Journal of the Linnean Society, 80(2), 353–368.

Grognet, P., Bidard, F., Kuchly, C., Tong, L. C. H., Coppin, E., Benkhali, J. A., Couloux, A., Wincker, P., Debuchy, R., & Silar, P. (2014). Maintaining two mating types: Structure of the mating type locus and its role in heterokaryosis in Podospora anserina. Genetics, 197(1), 421–432.
Hershey, A. D., & Chase, M. (2017). Independent Functions of Viral Protein and Nucleic Acid in Growth of Bacteriophage. Springer.
Holliday, R., & Grigg, G. (1993). DNA methylation and mutation. Mutation Research/Fundamental and Molecular Mechanisms of Mutagenesis, 285(1), 61–67.
Huberman, J. A., & Riggs, A. D. (1968). On the mechanism of DNA replication in mammalian chromosomes. Journal of Molecular Biology, 32(2), 327–341.
Hutchinson, G. E. (1979). The Kindly Fruits of the Earth: Recollections of an Embryo Ecologist. Yale University Press.
Ironside, J. E. (2010). No amicable divorce? Challenging the notion that sexual antagonism drives sex chromosome evolution. BioEssays, 32(8), 718–726.
Jaenisch, R., & Bird, A. (2003). Epigenetic regulation of gene expression: How the genome integrates intrinsic and environmental signals. Nature Genetics, 33(3), 245–254.
Kossel, A. (1913). Beziehungen der Chemie zur Physiologie. Die Kultur der Gegenwart; Teil 3, Abt. 3, Bd. 2: Mathematik, Naturwissenschaften, Medizin: Anorganische Naturwissenschaften: Chemie(4).
Kuroiwa, A., Ishiguchi, Y., Yamada, F., Shintaro, A., & Matsuda, Y. (2010). The process of a Y-loss event in an XO/XO mammal, the Ryukyu spiny rat. Chromosoma, 119, 519–526.
Lahn, B. T., & Page, D. C. (1999). Four evolutionary strata on the human X chromosome. Science, 286(5441), 964–967.
Mayr, C., Helm, K., Jakab, M., Ritter, M., Shrestha, R., Makaju, R., Wagner, A., Pichler, M., Beyreis, M., & Staettner, S. (2018). The histone methyltransferase G9a: A new therapeutic target in biliary tract cancer. Human Pathology, 72, 117–126.
McClung, C. E. (1902). The accessory chromosome—sex determinant? The Biological Bulletin, 3(1, 2), 43–84.
Mendel, G. (1865). Versuche über Pflanzen-Hybriden. Verhandlungen des Naturforschenden Vereins Brünn, 4, 3–47.
Meselson, M., & Stahl, F. W. (1958). The replication of DNA in Escherichia coli. Proceedings of the National Academy of Sciences, 44(7), 671–682.
Meselson, M., & Stahl, F. W. (1958). The replication of DNA. Cold Spring Harbor Symposia on Quantitative Biology.
Metzger, D. C., & Schulte, P. M. (2018). The DNA methylation landscape of stickleback reveals patterns of sex chromosome evolution and effects of environmental salinity. Genome Biology and Evolution, 10(3), 775–785.
Miescher, J. F. (1871). Ueber die chemische Zusammensetzung der Eiterzellen. https://www.abebooks.com/first-edition/chemische-Zusammensetzung-Eiterzellen-Dr-F-Miescher/19833462433/bd (accessed on 18 March 2025).
Mirsky, A. E. (1968). The discovery of DNA. Scientific American, 218(6), 78–90.
Morgan, T. H. (1910). Sex-limited inheritance in Drosophila. Science, 32(812), 120–122.
Morgan, T. H. (1913). Heredity and Sex. Columbia University Press.
Morgan, T. H. (1915). The Mechanism of Mendelian Heredity. H. Holt and Company.
Morgan, T. H. (1919). The Physical Basis of Heredity. J. B. Lippincott.
Morgan, T. H. (1926). The Theory of the Gene (Vol. 19). Yale University Press.

Morgan, T. H., Sturtevant, A. H., Muller, H. J., & Bridges, C. B. (1923). The Mechanism of Mendelian Heredity. H. Holt and Company.
Muller, H. J., & Painter, T. S. (1929). The cytological expression of changes in gene alignment produced by X-rays in Drosophila. The American Naturalist, 63(686), 193–200.
Myers, S., Bowden, R., Tumian, A., Bontrop, R. E., Freeman, C., MacFie, T. S., McVean, G., & Donnelly, P. (2010). Drive against hotspot motifs in primates implicates the PRDM9 gene in meiotic recombination. Science, 327(5967), 876–879.
Nieuwenhuis, B. P., & James, T. Y. (2016). The frequency of sex in fungi. Philosophical Transactions of the Royal Society B: Biological Sciences, 371(1706), 20150540.
Olby, R. C. (1994). The Path to the Double Helix: The Discovery of DNA. Courier Corporation.
Ot, A. (1944). Studies on the chemical nature of the substance inducing transformation of pneumococcal type. Journal of Experimental Medicine, 79, 137–158.
Painter, T. S. (1933). A new method for the study of chromosome rearrangements and the plotting of chromosome maps. Science, 78(2034), 585–586.
Painter, T. S. (1934). A new method for the study of chromosome aberrations and the plotting of chromosome maps in Drosophila melanogaster. Genetics, 19(3), 175.
Pokorny, J., & Smith, V. C. (1977). Evaluation of single-pigment shift model of anomalous trichromacy. Journal of the Optical Society of America, 67(9), 1196–1209.
Ponnikas, S., Sigeman, H., Abbott, J. K., & Hansson, B. (2018). Why do sex chromosomes stop recombining? Trends in Genetics, 34(7), 492–503.
Portin, P. (1993). The concept of the gene: Short history and present status. The Quarterly Review of Biology, 68(2), 173–223.
Portin, P. (2009). The elusive concept of the gene. Hereditas, 146(3), 112–117.
Portin, P. (2014). The birth and development of the DNA theory of inheritance: Sixty years since the discovery of the structure of DNA. Journal of Genetics, 93, 293–302.
Portugal, F. H., & Cohen, J. S. (1977). A Century of DNA: A History of the Discovery of the Structure and Function of the Genetic Substance. MIT Press.
Reich, D., Green, R. E., Kircher, M., Krause, J., Patterson, N., Durand, E. Y., Viola, B., Briggs, A. W., Stenzel, U., & Johnson, P. L. (2010). Genetic history of an archaic hominin group from Denisova Cave in Siberia. Nature, 468(7327), 1053–1060.
Reichwald, K., Petzold, A., Koch, P., Downie, B. R., Hartmann, N., Pietsch, S., Baumgart, M., Chalopin, D., Felder, M., & Bens, M. (2015). Insights into sex chromosome evolution and aging from the genome of a short-lived fish. Cell, 163(6), 1527–1538.
Rushton, A. R. (2014). William Bateson and the chromosome theory of heredity: A reappraisal. The British Journal for the History of Science, 47(1), 147–171.
Rushton, A. R. (2022). Cambridge geneticists and the chromosome theory of inheritance: William Bateson, Leonard Doncaster and Reginald Punnett 1879–1940. Annals of Science, 79(4), 468–496.
Southern, E. M. (1975). Detection of specific sequences among DNA fragments separated by gel electrophoresis. Journal of Molecular Biology, 98(3), 503–517.
Srb, A., & Horowitz, N. (1944). The ornithine cycle in Neurospora and its genetic control. Journal of Biological Chemistry, 154(1), 129–139.
Stern, K. (1931). Zytologisch-genetische Untersuchungen als Beweise für die Morgansche Theorie des Faktorenaustauschs. Biologisches Zentralblatt, 51(10), 547–587.

Sturtevant, A. H. (1913). The linear arrangement of six sex-linked factors in Drosophila, as shown by their mode of association. Journal of Experimental Zoology, 14(1), 43–59.
Sturtevant, A. H. (1914). The reduplication hypothesis as applied to Drosophila. The American Naturalist, 48(573), 535–549.
Subramanian, G., Adams, M. D., Venter, J. C., & Broder, S. (2001). Implications of the human genome for understanding human biology and medicine. JAMA, 286(18), 2296–2307.
Sutton, W. S. (1903). The chromosomes in heredity. The Biological Bulletin, 4(5), 231–250.
Watson, J. D. (1968). The text of The Double Helix: A personal account of the discovery of the structure of DNA. The Double Helix: A Personal Account of the Discovery of the Structure of DNA, 1–133.
Watson, J. D., & Crick, F. H. (1953). Molecular structure of nucleic acids: A structure for deoxyribose nucleic acid. Nature, 171(4356), 737–738.
Watson, K. M. (1954). Some general relations between the photoproduction and scattering of π mesons. Physical Review, 95(1), 228.
Wilkins, M. (1956). Physical studies of the molecular structure of deoxyribose nucleic acid and nucleoprotein. Cold Spring Harbor Symposia on Quantitative Biology, 21, 75–90.
Wilson, E. B. (1906). Studies on chromosomes III. The sexual differences of the chromosome-groups in Hemiptera, with some considerations on the determination and inheritance of sex. The Journal of Experimental Zoology, 3, 1.
Wittmann, H. G., & Wittmann-Liebold, B. (1966). Protein chemical studies of two RNA viruses and their mutants. Cold Spring Harbor Symposia on Quantitative Biology. 31, 163–172
Wright, A. E., Dean, R., Zimmer, F., & Mank, J. E. (2016). How to make a sex chromosome. Nature Communications, 7, 12087.
Xu, L., Auer, G., Peona, V., Suh, A., Deng, Y., Feng, S., Zhang, G., Blom, M. P., Christidis, L., & Prost, S. (2019). Dynamic evolutionary history and gene content of sex chromosomes across diverse songbirds. Nature Ecology & Evolution, 3(5), 834–844.
Yanofsky, C. (1963). Amino acid replacements associated with mutation and recombination in the A gene and their relationship to in vitro coding data. Cold Spring Harbor Symposia on Quantitative Biology, 28, 581–588.
Yanofsky, C., Ito, J., & Horn, V. (1966). Amino acid replacements and the genetic code. Cold Spring Harbor Symposia on Quantitative Biology, 31, 151–162.
Zamudio, N., Barau, J., Teissandier, A., Walter, M., Borsos, M., Servant, N., & Bourc'his, D. (2015). DNA methylation restrains transposons from adopting a chromatin signature permissive for meiotic recombination. Genes & Development, 29(12), 1256–1270.
Zhang, W., Wang, X., Yu, Q., Ming, R., & Jiang, J. (2008). DNA methylation and heterochromatinization in the male-specific region of the primitive Y chromosome of papaya. Genome Research, 18(12), 1938–1943.

CHAPTER 3

The Chromosome

UMBREEN BIBI, SARA ZAFAR,* and FIZA TALAT

Government College University, Faisalabad, Pakistan

*Corresponding author

ABSTRACT

The gene-carrying, thread-like structures found in a cell's nucleus are called chromosomes. Protein and deoxyribonucleic acid (DNA) make up the majority of a chromosome. Under a light microscope, they can be seen in a cell during the mitotic metaphase. The nuclear constituents of unique organization, individuality, and function are the chromosomes. Due to their ability to reproduce themselves, they are essential to the species' evolutionary progress, heredity, mutation, and variety.

3.1 HISTORICAL BACKGROUND

In the nucleus of plant cells, rod-shaped chromosomes were observed by Karl Nagli in 1842. The first comprehensive attempt to define chromosomes was made by E. Russow in 1872. One of the most important papers on the relationship between chromosomes and cell division (CD) stages was written by A. Schneider in 1873. The discovery of thread-like filaments that emerged during CD was made by E. Strasburger in 1875. The term "chromatin" was first used by Walter Flemming in 1878 to refer to the nucleus's thread-like substance that took on a vibrant color following staining. In 1883, W. Roux had concerns about the chromosomes' role in inheritance. According to Benden and Boveri (1887), each species has the same number of chromosomes. W. Waldeyer (1888) gave the term "chromosome" (Gr.,

Chromosomal and Genomic Structures and Functions,Volume 1: Chromosomes and Cell Division.
Tariq Ahmad Bhat and Jameel M. Al-Khayri (Eds)

DOI: 10.1201/9781779640253-3

chrom = color, soma = body) to refer to the nucleus's darkly stained bodies. In 1902, W. S. Sutton and T. Boveri proposed that the physical structures known as chromosomes served as the transmitters of heredity. One maternal and one paternal member make up each chromosomal pair in a synapsis, according to Sutton's (1902) observation. He thought that the Mendelian rules of inheritance might have a solid base due to the way chromosomes function. He is recognized as the creator of the chromosomal foundation of heredity hypothesis (Verma and Agarwal, 1999).

The cytological foundation for the rules of heredity was developed in the early 1900s by Thomas Morgan and Hermann Müller. In 1922, using *Drosophila* chromosomes, 2000 genetic variables were found on the four chromosomes of the fruit fly. Robert Feulgen introduced the Feulgen reaction, a color test for DNA, in 1914. He demonstrated that chromosomes comprise DNA in 1924. Brachet proved the existence of another nucleic acid, RNA, in 1942 by utilizing cytochemical techniques. Shortly after, Mirsky and Pollister (1946) revealed the existence of proteins connected to chromosomal material. Kaufmann (1948) provided a description of chromosomal morphology (Goldschmidt, 2022). According to Dupraw's (1965) "folded fiber model," a single, tightly folded DNA molecule covered with chromosomal proteins makes up a chromosome. The fundamental chromatin material was proposed as the "nucleosome model" by R. D. Kornberg in 1974. P. Outdet and colleagues first used the word "nucleosome" in 1975 (Van Holde, 2012).

3.2 CHROMOSOME STRUCTURE

3.2.1 CENTROMERE

The centromere is a small, constricted area of the chromonema that is believed to be a permanent part of the chromosome.

- The two chromonemata have now been connected.
- The chromosome is divided into two portions, or "arms," by the centromere, kinetochore, or main constriction. The "p arm" refers to the short arm of the chromosome. The term "q arm" refers to the chromosome's long arm.
- Its location is fixed for a specific type of chromosome and serves as an identifying characteristic.

- During CD, the spindle fibers at this location are where the chromosomes are attached.
- Thin electron microscopic sections of the kinetochore reveal a trilaminar structure, consisting of a dense outer proteinaceous layer 10 nm thick, a low-density middle layer, and a dense inner layer closely attached to the centromere (McKinley et al., 2016).

3.2.1.1 NUMBER OF CENTROMERE

1. **Acentric Chromosome:** An acentric chromosome is one that lacks a centromere. During CD, this type of chromosome remains weak and finally disappears.

2. **Monocentric Chromosome:** Monocentric chromosomes are defined as having a single centromere. It represents a typical sort of chromosome.

3. **Dicentric Chromosome:** A dicentric chromosome is a chromosome that has two centromeres. These chromosomes are created by translocations and inversions, which result in a dicentric bridge during anaphase.

4. **Polycentric Chromosome:** A chromosome with two or more centromeres (Padeganeh et al., 2013).

3.2.1.2 ARM

Any segment of a chromosome that is between a centromere and one extremity. The length of arms with secondary constrictions may vary. This phenomenon is related to the fluctuations in the secondary constriction's despiralization. The appropriate length is regarded as the average length (Battaglia, 1955).

1. **Median Centromere:** A centromere that is positioned in the center of the chromosome, giving it a 1:1 arm ratio.

   i. **Submedian Centromere:** A centromere that is close to the center of the chromosome, giving rise to an arm ratio that is more than 1:1 but lower than 1:3.

ii. **Subterminal Centromere:** A centromere close to one chromosomal extremity that causes an arm ratio of at least one to three.

2. **Terminal Centromere:** A centromere that is located at one end of the chromosome, giving it a 0:1 arm ratio.

3. **Isobrachial Chromosome:** Median centromere chromosome

4. **Heterobrachial Chromosome:** Sub: median centromere chromosome

5. **Hyper Heterobrachial Chromosome:** Subterminal centromere chromosome (Battaglia, 1952).

6. **Monobrachial Chromosome:** Terminal centromere chromosome (Sorokin, 1929).

3.2.2 SATELLITE

E chromosomal segment distant from a nucleolar constriction. Each satellite has a single nucleolus at telophase. Nevertheless, rather than the satellite itself, the nucleolar constriction is connected to nucleolus development. Therefore, based on the above definition, this interpretation implies that the satellite is a section of the chromosome that is externally near a nucleolar constriction. A satellite is traditionally thought to be a spheroidal body whose diameter is either smaller than or equal to that of a chromosome. It is attached to the chromosomal body at a single point by a fine thread. Nonetheless, the nucleolar constriction location alone is responsible for that appearance. Consequently, the satellite is longer than it is wide when the nucleolar constriction is distant from the chromosome extremity. It becomes apparent as an extended chromosomal segment and loses its distinctive "spheroidal shape" (Cechova et al., 2022).

1. **Microsatellite:** A tiny spheroidal satellite, defined as one whose diameter is half or less of the chromosomal diameter.

2. **Macro Satellite:** A large-sized spheroidal satellite, defined as one whose diameter is more than half of that of a chromosome.

3. **Linear Satellite:** A satellite shaped like a lengthy chromosomal section, additionally, based on the satellite's position.

4. **Terminal Satellite:** A satellite situated between one extremity and its nucleolar constriction.

5. **Intercalary Satellite:** Two nucleolar constrictions are separated by a satellite. A terminal satellite and an intercalary satellite are typically illustrated by the well-known tandem satellites. It is evident from cytological examinations that nucleolar constrictions take many different shapes, from those that are reduced to a thin thread to those that have the potential to reach the diameter of the chromosomal body. Chromosomes that include one or more satellites could thus be referred to as satellite chromosomes (Liao et al., 2023).

3.2.3 PSEUDO SATELLITE

A spherical region of the chromosome that is next to the centromere. A chromosome's short arm with a subterminal centromere serves as a more prevalent example of a pseudo satellite. This instance involves a terminal pseudo satellite. The existence of a nucleolar constriction shifts the pseudo satellite's position from terminal to intercalary in an indirect manner. As a result, an intercalary pseudo satellite is created.

3.2.4 CHROMATID

Each chromosome has two symmetrical structures called chromatids as it enters mitotic metaphase. There is one DNA molecule per chromatid. When a chromosome's sister chromatids migrate to opposite poles at the start of anaphase, the centromere is the only thing holding the two chromatids together (Peters et al., 2012).

3.2.5 CHROMONEMATA

The chromosomal material is seen as extremely thin filaments known as chromonemata during mitotic prophase. Early in the condensation process, a chromatid is represented as a chromonema. As a result, the terms "chromatid" and "chromonema" describe the same structure, which consists of a single linear DNA molecule and the proteins that bind to it. Chromosome segments that contain genes are referred to as chromonemata. A chromosome can contain many chromonemata, each of which is embedded in the amorphous, achromatic matrix. There is a pellicle or sheath around the matrix. Non-genetic elements like the matrix and pellicle only emerge during metaphase, when the nucleolus vanishes. It is believed that the matrix and nucleolar substances are interchangeable; that is, the nucleolus arises when the chromosomal matrix does, and vice versa. Nevertheless, studies made using electron microscopy show the presence of pellicle and matrix in them (Câmara et al., 2023).

3.2.6 CHROMOMERES

When chromosomes are in interphase, they can be seen as bead-like buildup of chromatin material called chromomere-bearing chromatin, which resembles a necklace with multiple beads on a string. In polytene chromosomes, the chromomeres become particularly evident when they align side by side to form chromosome beads. When chromosomes are in metaphase, they are tightly wrapped and the chromomeres are hidden. Cell biologists are particularly interested in chromomeres, which are sections of DNA that are tightly folded. They are thought to line up with the chromosomal units of genetic function. For a very long time, the majority of geneticists believed that these chromomeres were genes—that is, the basic units of heredity (Maeshima et al., 2019).

3.2.7 TELOMERE

(Greek: meros = part; telo = for) Each chromosome extremity has a polarity; it prevents the fusion of other chromosomal regions with it. The telomeres are the ends of the chromosomes, and because telomeres are absent from broken ends of chromosomes, they can fuse together (Bailey et al., 2006).

3.2.8 SECONDARY CONSTRICTION

The chromosomes have secondary constriction at any location along the chromosome in addition to the primary constriction and the centromere. Since these constrictions are consistent in their location and extent, they can be useful in identifying certain chromosomes within a set. Since the centromere is the sole location on the chromosome where the chromosome bends during anaphase, secondary constrictions can be separated from primary constrictions or centromeres (Palmer et al., 2004).

3.2.9 NUCLEOLAR ORGANIZERS

These are specific secondary constrictions containing genes encoding 5.8S, 18S, and 28S ribosomal RNA, which stimulate nucleoli development. The reason for the secondary constriction could be that the rRNA genes interfere with chromosomal condensation by being actively translated. The nucleolar organizers in humans are found in the secondary constrictions of chromosomes 13, 14, 15, 21, and 22, which are all acrocentric and have satellites (Alberts et al., 2002).

3.3 CHROMOSOME SHAPE

3.3.1 V SHAPE CHROMOSOME

A chromosome is referred to as a V-shaped chromosome when it takes on a V shape during the anaphase stage. Among them is the metacentric chromosome.

3.3.2 J SHAPE CHROMOSOME

J-shaped chromosomes are defined as chromosomes that take on the J shape during the anaphase stage. Subterminal and submetacentric chromosomes are among them.

3.3.3 ROD SHAPE CHROMOSOME

A chromosome is said to have a rod-like form when it enters the anaphase stage. The telocentric chromosome is among them.

3.3.4 *LINEAR CHROMOSOME*

A chromosome is referred to as linear if it has a linear structure or if both ends are free. Eukaryotes have these kinds of chromosomes.

3.3.5 *CIRCULAR CHROMOSOME*

A circular chromosome is a chromosome that has a circular structure and form. Both viruses and bacteria contain them.

3.4 CHEMICAL COMPOSITION

Protein, RNA, and DNA can all be found in chromatin that has been extracted from rat liver. There are two different kinds of chromatin proteins: histones and non-histones. Chromatin has been modeled using rat liver chromatin. It has an approximate 1:1 histone to DNA ratio, a 0.6 non-histone protein to DNA ratio, and a 0.1 RNA/DNA ratio.

3.4.1 *DNA*

As DNA is essential for regulating heredity, it is the most significant chemical constituent of chromatin. The picogram (10–12 grams) is the most practical unit of measurement for DNA. The two phenomena represented by chromatin DNA are as follows:

The Feulgen reactions were used to stain the DNA in nuclei, and a cytophotometer—a specialized microscope—was used to detect the stain concentration in individual nuclei. The methods showed that the amount of DNA in nuclei is always the same. Therefore, if an organism's cells are diploid, they all have the same amount of DNA (2C). Due to their haploid nature, gametes only contain half as much DNA (1C). There are occasionally polyploid cells in some tissues, such as the liver, and these cells' nuclei contain more DNA as a result.

3.4.2 *HISTONES*

Because of the high concentration of the amino acids arginine and lysine—roughly 24 moles per protein—histones are extremely basic proteins.

Arginine and lysine are cationic at physiological pH and can interact electrostatically with anionic nucleic acids. The basic histones have a strong binding with the acidic DNA. The eukaryotic chromosomes include five different forms of histones: H1, H2A, H2B, H3, and H4. One of the important discoveries to come from chemical studies is the remarkable preservation of the basic structures of histones throughout evolutionary history. For instance, there are only two amino acid changes between the calf and garden pea histone H4, which is a protein with 102 residues (DeLange, 1969). Similarly, out of 102 total amino acid residues, the sequence of histone H3 in rats differs from that in peas by just two amino acids. These species are thought to have undergone structural divergence during the course of at least 600 million years of evolution. This structural conservation demonstrates the similar and significant role of histones in maintaining the structural and functional integrity of chromatin. These two histones' roles presumably entail almost all of their amino acids due to their evolutionary conservation; any alteration to one of their positions would be harmful to the cell.

3.4.3 NON-HISTONES

Non-histone proteins are more diverse than the small population of histones found in chromatin. The number of non-histones might range from 12 to 20 in different organisms. Unlike histones, these proteins' heterogeneity has not been preserved throughout evolution. These non-histones vary even within the same organism's tissues, indicating that they control the expression of particular genes. Actin, myosin, and the α- and β-tubulins are among the approximately 50% of non-histones in chromatin that are structural proteins. These contractile proteins are now recognized as vital parts of the chromosome, acting as chromosomal movers during mitosis, meiosis, and chromosome condensation. Previously, it was thought that these proteins were contaminants (Kinoshita and Hirano, 2017). Largely, the remaining 50% of non-histones are made up of all the enzymes and other components involved in transcription, DNA replication, and transcription regulation. While these proteins must perform similar enzymatic functions, they are not as extensively conserved across organisms. They don't seem to be as crucial for preserving chromosomal integrity as histones are (Shvedunova et al., 2022).

3.5 ULTRASTRUCTURE

It is not clearly understood how DNA is organized in the chromatin using electron microscope technology in the domain of ultrastructure of the chromatin. Both complete chromosome mounts and individual chromosome segments were examined for chromosomal research under an electron microscope. Research has shown that chromosomes consist of minuscule fibrils, with a thickness ranging from 2 to 4 nm.

3.5.1 SINGLE-STRANDED AND MULTI-STRANDED HYPOTHESES

One of two reasons for the observed variations in DNA content between related species can be identified when comparing their chromosomes: one of two processes can occur: (i) chromonemata multiply laterally, resulting in multiple or multi-strandedness; and (ii) tandem duplication of DNA, also known as chromonemata, in which lengthwise duplication results in chromatin polymorphisms. The latter circumstance will keep chromosomes single-stranded. While some examples involving plants, such as *Vicia faba*, and animals, such as dipteran salivary gland chromosomes, have demonstrated multiple strandedness, there are indications that prevent this theory from being applied universally. However, it is clear that tandem duplication of chromonemata, or DNA, occurs in each of these instances. The concept that chromatin is single-stranded is, in fact, supported by an extensive amount of evidence. A single linear DNA molecule is used to create each chromosome in the yeast *Saccharomyces cerevisiae*, as the pulsed gel electrophoresis technique shows (Kavenoff et al., 1974).

3.5.2 FOLDED-FIBER MODEL AND NUCLEOSOME

DNA should be present in a coiled or folded form if we assume that each chromatid contains a single lengthy DNA molecule. There was disagreement over the way that DNA is coiled and folded, and there were several models available; just two stood out as being very significant. The folded-fiber model represents a well-liked design (Dupraw, 1965). It claims that a tightly folded fiber with a reasonably uniform diameter of 200 to 300 Å makes up the majority of the chromosome. The DNA-histone helix (with a diameter of 30 Å) is expected to be supercoiled within this folded fiber. The nucleosome model is the most significant and well-known model. Consequently, the nucleosome model has supported the opposite of the folded-fiber concept, which proposed that the histones are bonded on the outside of the DNA coils (i.e., a histone shell around DNA), that is, a histone particle with DNA surrounding it. To put it another way, the previous hypothesis (Berns et al., 1983) that basic chromatin fiber contained a DNA core surrounded by histones was untrue. In actuality, the topology of the packing process through the nucleosomes is important from a genetic standpoint because it exposes the DNA freely across the entire "spool" surface, allowing for genetic expression, rather than being buried at any point. It appears that nucleosomes are a ubiquitous tool for compacting eukaryotic cells' lengthy DNA strands (Hirano, 2000).

3.5.3 NUCLEOSOMES AND SOLENOID MODEL OF CHROMATIN

In eukaryotes, nucleosomes—a recurring collection of DNA-protein particles—are formed by the tight binding of DNA to an equivalent mass of histones. The double helix of DNA in every human chromosome could reach thousands of times across the cell nucleus. Histones are necessary to neatly compress this incredibly long DNA molecule—as a nucleosome—into a nucleus that is only a few micrometers in diameter. Therefore, the fundamental packing units of chromatin are called nucleosomes, which give chromatin the appearance of "beads on a string" in electron micrographs produced by techniques that highlight higher-order packing (Olins and Olins, 1974). The lengthy DNA "string" can have its nucleosome "beads" removed by breaking it down with DNA-degrading enzymes like the bacterial enzyme micrococcal nuclease. Only the DNA among the nucleosome beads is broken down by micrococcal nuclease after a brief period of digestion. The remaining portion is resistant to digestion and persists as 146 nucleotide-long stretches of double-stranded DNA attached to a particular complex of eight nucleosome histones (the histone octamer). The nucleosome beads that were produced in this manner were crystallized and subjected to X-ray diffraction analysis. H1 nucleosome in DNA. With a diameter of around 11 nm and a height of 5.7 nm, the four nucleosome histones, H2A, H2B, H3, and H4, are found in two copies in each disc-shaped nucleosome. The double-stranded DNA helix, which has 146 base pairs, is wrapped once around the protein core created by this histone octamer, which is a core of histone tetramer H3, H4, and the polar parts of dimer H2A and H2B. The DNA runs continuously from nucleosome to nucleosome. A segment of linker DNA, typically spanning 54 base pairs and containing a single H1 histone protein molecule, divides each nucleosome from the next. Typically, DNA completes two turns around the histone octamers, with H1 molecules severing the two 200 bp long twists (Reid et al., 1980). As a result, nucleosomes typically repeat every 200 nucleotides or base pairs. For instance, each human cell containing 6×10^9 DNA nucleotide pairs contains 3×10^7 nucleosomes, and a eukaryotic gene with 10,000 nucleotide pairs will be linked to 50 nucleosomes.

3.6 MATERIALS OF CHROMOSOMES

Based on the proportion of DNA, RNA, and proteins in the material and how well it stained, the chromatin material of eukaryotic chromosomes has been divided into the following categories.

3.6.1 EUCHROMATIN

The extended form of chromatin, known as euchromatin, makes up the majority of chromosomes. Because its constituent DNA molecules can only synthesize RNA molecules in the stretched form of chromatin, it has a unique affinity for basic stains and is biologically active (Adam et al., 2001).

3.6.2 HETEROCHROMATIN

A condensed, inter-coiled form of chromatin is called heterochromatin. It has two to three times as much DNA as euchromatin. Since it does not directly synthesize RNA (transcription) or protein, it is genetically inert. Moreover, it frequently replicates at a different rate than the rest of the DNA (Kosak et al., 2002).

Three different types of heterochromatins have recently been identified, including:

1. Constitutive.
2. Facultative.
3. Condensed heterochromatin.

1. **Constitutive:** Constitutive heterochromatin is found in the nucleus of almost every cell in an organism and is present constantly. It tends to aggregate into chromocenters or false nucleoli in an interphase nucleus.

   Highly repetitious satellite DNA found in constitutive heterochromatin replicates late; it doesn't start to replicate until late in the S-phase and then resumes for only a short period just prior to the G2 phase (Bernard et al., 2001).

2. **Facultative:** The facultative heterochromatin is indicative of the presence of a regulatory mechanism intended to modify the nucleus's "dosages" of specific genes.

   It is arranged during a process known as facultative heterochromatin, in which some chromosomes become euchromatic in cells of the opposing sex and heterochromatic in cells of the same sex. It is still somewhat connected to sexual differences (Feil, 2007).

3. **Condensed Heterochromatin:** It is a particular kind of highly stained, tightly coiled chromatin that is seen in many interphase nuclei and has a specific function in gene regulation. It is distinct from the other two varieties of chromatin.

3.7 TYPES OF CHROMOSOMES

3.7.1 ALLOSOMES

Genes that define sex are carried by chromosomes that are different in shape and number in male and female sexes. They can be divided into two groups, generally speaking: Z and W types or X and Y types.

3.7.2 AUTOSOMES

Chromosomes that are similar in size and shape in males and females, and which are typically unrelated to sex-determining genes (Miller and Therman, 2011).

3.7.3 SPECIAL TYPES OF CHROMOSOMES

In certain cells or at specific stages of their life cycle, eukaryotes have unique and special types of chromosomes in addition to the common types (autosomes and sex chromosomes). These unique eukaryotic chromosomal types are:

1. Polytene Chromosome.
2. Lampbrush Chromosome.
3. B Chromosome.
4. Holokinetic Chromosome.

3.7.3.1 POLYTENE CHROMOSOME

The exceptionally long and wide chromosomes seen in the nucleus of the salivary gland cells of dipteran larvae, such as *Drosophila*, are known as polytene chromosomes. These chromosomes are approximately 100–200 times larger than those in the same species' meiosis and mitosis. The process of "endomitosis," in which salivary gland cells do not split after the glands are established but instead have their chromosomes replicated numerous times, causes the chromosomes to grow to an unusually large size. Another name for them is multistranded chromosomes. 2X chromosomes are produced as a result of endomitosis, where X indicates the number of multiplications.

3.7.3.1.1 *Polytene chromosomes of D. melanogaster*

It is made up of 1000–2000 chromosomes that are the result of 9–10 continuous cycles of multiplication. They have bands of light and dark color that alternate along their length. Like the chromomeres of a basic chromosome, the black bands are disc-shaped structures that encircle the entire chromosome. They contain euchromatin. Heterochromatin makes up the fibrillar light bands, also known as interbands. This chromosome also has a huge region known as a "puff," which moves depending on the stage at which the larvae are developing. This reflects changes in gene activity and includes a number of processes, including the synthesis of mRNA and the formation of chromosomal loops (Roy et al., 2021).

3.7.3.2 LAMPBRUSH CHROMOSOME

Several chromosomes with a lampbrush form and extremely large nuclei are found in the yolk-rich oocytes of vertebrates during the diplotene stage of meiosis.

- The lampbrush chromosomes are created during the active production of mRNA molecules for use by the egg during cleavage, which occurs when chromosome involvement in mitotic CD prevents mRNA molecule synthesis.
- The main axis of a lampbrush chromosome is made up of chromosomal fiber, or DNA molecules, which branch out laterally along its length.
- The loop generates several types of m-RNA molecules.
- In a mature egg, the lateral loops disappear when the chromosome contracts (Krasikova et al., 2023).

3.7.3.3 B-CHROMOSOME

- The B-chromosome, also known as supernumerary chromosomes, accessory chromosomes, accessory sections, and so forth, is a unique form of chromosome found in many plant and animal species that has no discernible genetic purpose.
- They can be primarily euchromatic or predominately heterochromatic (insects, maize, etc.).
- In animals, B chromosomes are exclusively preserved in the cell lines that give rise to the reproductive organs; they vanish from somatic, or non-reproductive, tissue.
- B-chromosomes are harmful to the organism because they generate aberrant nucleus divisions in gametophyte plants and inappropriate crossing over during animal meiosis.
- It is unknown where the B-chromosomes originated. Although it is not always the case, some animals may have sex chromosome derivatives.
- It is not paired with A chromosome (Ahmad et al., 2019).

3.7.3.4 HOLOKINETIC CHROMOSOME

Centromeres are located at a precise location on each chromosome in the majority of plants and animals. However, the kinetic activity is spread throughout the entire chromosome in a few plants and a number of animals, including insects. These chromosomes are known as holokinetic chromosomes.

Holokinetic chromosomes are those in which the kinetic activity is spread throughout the entire chromosome in a variety of species, particularly in insects of the order Hemiptera and a few largely monocotyledonous plants (Juncales, Cyperales) (Sybenga, 2000).

Although it is not entirely rational, the term "diffuse centromere" has been employed as a substitute. A holokinetic chromosome's chromatids align parallel to one another at the equator during mitotic metaphase, with one chromatid facing one pole and the other facing the opposite pole. They also split in this manner during anaphase, and they continue in this direction until they reach the poles. Kinetic activity most likely begins at one location and continues from there, correlating each unit with the one before it (Mandrioli et al., 2020).

3.7.4 HOMOLOGOUS CHROMOSOMES

A homologous chromosome is a pair of chromosomes that are coupled during meiosis to produce sex cells. These chromosomes are similar in terms of centromere location and size. Additionally, they share the same genes at the same location, though they may not have the same alleles.

3.8 KARYOTYPE AND IDIOGRAM

A pair of chromosomes with shared fixed features identifies each individual within a species of plant or animal. The number of chromosomes, their proportional sizes, the centromere's location, the arms' lengths, secondary constrictions, and satellites are some of these features. The collection of traits that designate a specific set of chromosomes is known as a karyotype (Heslop and Schwarzacher, 2011). An idiogram is a graphic representation of the karyotype, or chromosomal morphology, of a species (Greek: idios = distinctive, gramma = anything written). In an idiogram, an organism's haploid set of chromosomes is often arranged in a series of progressively smaller sizes. An idiogram, which arranges homologous pairs in decreasing order of size, can occasionally be created for a diploid set of chromosomes (Gasparotto et al., 2022). An image displays a group of mitotic chromosomes. They can be used to identify homologous chromosomal pairs and to prepare karyotypes. Their microphotographs provide a karyotype of the human metaphase chromosomes. After being extracted from the microphotographs, each chromosome is positioned in size-related alignments with its corresponding partner. Finding the "centromeric index," or the ratio of the chromosome's long to short arms' lengths, can help to improve the method. Certain species may have unique karyotypes; for instance, plants usually have heterochromatic areas at the telomeres, mice have acrocentric chromosomes, and many amphibians only have metacentric chromosomes. Karyotypes' uses: occasionally, the karyotypes of different animals are investigated, and resemblances between them are thought to indicate an evolutionary link. A karyotype can also allude to an organism's basic or sophisticated characteristics. Either symmetry or asymmetry could exist. An asymmetric karyotype is one that has fewer metacentric chromosomes and shows significant disparities between the set's smallest and largest chromosomes. *Ginkgo biloba* is an example of an asymmetric karyotype, which is thought to have a more advanced trait than a symmetric karyotype. There is a noticeable trend in flowering plants toward asymmetric karyotypes. In the family Compositae, this tendency has been well investigated. It has been demonstrated time and time again that specialized zygomorphic flowers are linked to enhanced karyotype asymmetry.

3.9 ISOLATION METHODS

Over time, a number of techniques have been devised to study chromosomes. Preparing pure nuclei from the liver, thymus, or any other material first is a somewhat easy method. After lysing the nuclei with detergent, the chromatin is pelleted in a preparative centrifuge at 20,000 rpm, or roughly 50,000 g. Biochemical isolation of chromatin can also be achieved by lysing nuclei in hypotonic solutions after they have been purified. Chromatin appears as a sticky, gelatinous substance when prepared in this manner. Squash preparations are the best for observing chromosomes.

Certain stains, like acetocarmine, Giemsa, or orcein, are applied to stain certain areas, such as tissue fragments like the tips of onion roots. Onion roots are stained with basic stains like acetocarmine, Giemsa, or orcein and then gently compressed between the coverslip and slide. Sometimes the nucleus swells, and the individual chromosomes are better separated by using hypotonic solutions before squashing. The optimum time to study chromosomal morphology is during metaphase and anaphase, when chromosome contraction is at its greatest (Mateuca et al., 2006).

3.10 FUNCTION AND GENETIC SIGNIFICANCE OF CHROMOSOMES

The number of chromosomes in a species is constant and therefore plays a crucial role in determining the species' taxonomy and phylogeny.

3.10.1 GENETIC CODE STORAGE

The chromosome contains the genetic material that an organism needs to grow and develop. DNA molecules are made up of a series of units called genes. Sections of DNA known as genes are responsible for encoding particular proteins that are necessary for the normal functioning of cells.

3.10.2 SEX DETERMINATION

Human chromosomes are found in 23 pairs, one of which is the sex chromosome. Males have one X and one Y chromosome, while females have two X chromosomes. It is the father's chromosome that determines the sex of the child. When an X chromosome is inherited from the father, the child will develop into a female, and *vice versa*.

3.10.3 CONTROL OF CELL DIVISION (CD)

During mitosis, chromosomes verify that cells divide successfully. The parent cell's chromosomes ensure that the daughter cells receive the proper information needed for appropriate cell growth and development.

3.10.4 PROTEIN FORMATION AND STORAGE

The chromosomes control how our bodies make proteins and keep DNA in its proper order. The coiled shape of the chromosomes stores the proteins as well. These DNA-attached proteins help ensure that DNA is packaged correctly (Alberts et al., 2002). Because they are the only connection between two generations, chromosomes are regarded as the organs of heredity. A diploid chromosomal set is made up of two sets of anatomically similar chromosomes, one from each parent at fertilization (excluding the X and Y chromosomes). The genetic material, which can be either DNA or RNA, is arranged in chromosomes and remains mostly unchanged from generation to generation.

The genetic information found in each DNA molecule is maintained and duplicated by the chromosomes. This information is transcribed into specific types of RNA molecules at the appropriate times and in the correct sequences, directing the synthesis of various protein types to form an organism that resembles its parents (Gutbrod et al., 2020).

3.11 PROKARYOTIC AND EUKARYOTIC CHROMOSOMES

3.11.1 CHROMOSOME NUMBER AND GENOME SIZE

The number of chromosomes in a species is constant. They therefore play a crucial role in determining the species' taxonomy and phylogeny. The term "reduced or haploid gametic sets of chromosomes" refers to the quantity or collection of chromosomes found in gametic cells, such as sperm and eggs. The genome is another name for the haploid collection of chromosomes. The majority of organisms' somatic, or body, cells are known as diploid cells because they have two haploid sets of genomes. Through the combination of the haploid male and female gametes during sexual reproduction, the diploid cells acquire the diploid set of chromosomes.

Chromosome "sets" are denoted with the suffix "ploid." The degree of ploidy is indicated by the prefix. Finally, "2n" represents the diploid or somatic chromosomal number in an individual, whereas "n" typically denotes the gametic or haploid chromosomal number. However, in polyploid individuals, the so-called base number, or ancestral primitive number, must be determined. This number is symbolized by the symbol "x." For instance, ordinary wheat is hexaploid ($2n = 6x$) in *Triticum aestivum,* where

2n = 42, n = 21, and x = 7 (Ozkan et al., 2003). It's common knowledge that prokaryotes have just one chromosome. However, prokaryotic cells' small size has typically made it impossible to count their chromosomes directly. The chromosomal sizes of different prokaryotic organisms vary. For example, the chromosomes of *Escherichia coli* and *Mycoplasma* are 100 and 265 micrometers long, respectively. In recent times, the number of chromosomes has been deduced by DNA fragment mapping and pulsed-field gel electrophoresis. These methods have resulted in an increasing number of organisms possessing multiple chromosomes, along with extremely large plasmids. Two large circular mapping chromosomes are found in Vibrio species, two in *Deinococcus radiodurans* (Yamaichi et al., 1999), three in *Rhizobium meliloti* isolates, four in Burkholderia (*Pseudomonas*) cepacia isolates (Lin et al., 1999), and three in *Rhodobacter sphaeroides*, *Leptospira interrogans*, and *Brucella* species (Lessie et al., 1996). According to Jumas et al. (1998), certain *Agrobacterium* species have one circular and one linear mapping chromosome, while *Paracoccus denitrificans* has three unmapped chromosomes (Winterstein et al., 1998). Since the ant *Myrmecia pilosula* has only one (2n=2) chromosome, prokaryotes can even have more chromosomes than eukaryotes (Crosland et al., 1986). Prokaryotes are capable of having bigger genomes. For instance, compared to the eukaryotes *Encephalitozoon* species (2.3 and 2.9 Mb) (Peyretaillade et al., 1998) and *Pneumocystis carinii* (8.8 Mb) (Bendich and Drlica, 2000), the prokaryotes *Myxococcus xanthus* (9.5 Mb of DNA) (Chen et al., 1991) and *Calothrix* strains (12 Mb) (Herdman et al., 1979) have larger genomes. The number of distinct genes in *M. Xanthus* is estimated assuming a density of one open reading frame per 1.1 kb of prokaryotic chromosomal DNA (Doolittle et al., 1998). It is clear that the existence or lack of a nuclear membrane is not absolutely correlated with either chromosomal number or genome size (or, more likely, gene number).

3.11.2 CHROMOSOME COPY NUMBER

Although ploidy may exceed 500 in some tissues, such as the salivary glands of certain insects and the endosperm of maize, metazoans and plants are primarily diploid (John, 1988). Prokaryotes, on the other hand, are believed to be genetically monoploid. However, compared with many animal and plant tissues, prokaryotes may have a larger copy number of genomic DNA. For instance, during early exponential development in a rich medium, an *Escherichia coli* cell has roughly 11 genome equivalents (with a range of roughly 5–18) and 2, 4, or 8 at stationary phase (Akerlund et al., 1995). In the stationary phase, *Methanococcus jannaschii* has an average chromosomal copy number of three (range 1±5), but during exponential growth, it is approximately seven (range 3±15) (Malandrin et al., 1999). Comparable chromosomal copy numbers are found in *D. radiodurans*, where four occur during stationary phase and 10 during exponential growth (Hansen, 1978).

3.11.3 CIRCULAR OR LINEAR FORM AND TELOMERES

The well-known autoradiographic image of a circular replicating *E. coli* chromosome that Cairns (Maldonado et al., 1994) presented seemed to validate previous genetic mapping data in this species and laid the groundwork for the theory that prokaryotic chromosomes are essentially distinct from the linear structures present in eukaryotes. In bacteria, however, linear chromosomes can exist. These bacterial chromosome ends also bear resemblance to eukaryotic chromosomal telomeres, which are structures engineered to address the "ends problem" in linear DNA replication (Casjens et al., 1997). For instance, *B. burgdorferi*'s chromosomal ends are hairpin formations. The ends of *Streptomyces* have a number of short, inverted, repeating sequences that can form hairpins. Proteins that are likely complementary to the three ends of the DNA are connected to the five termini of the fungus. Therefore, circular chromosomes are not found in all bacteria. The most frequent chromosomal form in prokaryotes may not even be the circle, since so few species have undergone thorough analysis. Moreover, the proof of circularity for the majority of bacterial chromosomes comes solely from mapping data, which is insufficient to differentiate between circular and linear chromosomes (the circular map of bacteriophage T4 DNA, for instance, originates from a circular permutation of linear DNA). Given that practically all of the molecules discovered by Cairns' method were confusing tangles and linear forms, even the circular form he discovered might not be representative (Cairns, 1962). It is believed that chromosomes are linear in eukaryotic cells. Nonetheless, under specific circumstances, it is possible to sustain only circular forms: the majority of telomerase-deficient fission yeast cells that survived did so by circularizing their chromosomes (Nakamura et al., 1998). Therefore, there is no assurance that the prokaryotic/eukaryotic form of cellular architecture and the circular/linear form of chromosomes are correlated. Differential chromatin and transcriptional silencing mean an organism does not use all of the genetic information it carries with it during its life cycle. Certain genomic areas are transcriptionally silenced by being compacted with proteins in order to ensure deactivation. The term "eukaryotic silencing" describes the suppression of gene expression that emanates from the silencer, a DNA regulatory region (Rine, 1999). Highly compacted heterochromatin production and silencing may be interchangeable concepts. *E. coli* also experiences silencing, which extends from the sopC site on the F plasmid and the parS site on plasmid P1 (Rodionov et al., 1999) over a distance of roughly 10 kb (Kim et al., 1999). Similar to eukaryotes, some proteins are needed for bacterial silencing (in this case, ParB and SopB in *E. coli*). If a compacted DNA-protein complex is big enough to bind enough Giemsa, Feulgen, or another cytological dye, it can be identified cytologically. This is the situation in large nuclear genome eukaryotes, where silent DNA is identifiable as heterochromatin, either as individual chromosomes or as compacted complete chromosomes. However, the genomes of other organisms, such as prokaryotes and baker's yeast, are too tiny or contain insufficient silenced DNA to allow for in situ staining detection. However, bacteria do exhibit a general shutdown of genomic activity that is associated with chromosomal DNA compaction. For instance, when *Bacillus* sporulates, its transcriptional activity stops and new proteins are coupled to the chromosome (Setlow, 1995). Comparably, *Chlamydia trachomatis* likely induces chromosomal condensation when it transforms from an inactive, extracellular elementary body form to a metabolically active reticulate body (Christiansen et al., 1993) due to a histone H1-like protein. When the archaebacterium, that is, *Halobacterium salinarium*, moves from the early to late exponential phase of growth, a third instance is seen. Nucleoid sedimentation qualities also reflect this transition, as the nucleoid changes from a form containing naked DNA to one with the beads-on-string form typical of nucleosomal DNA (Takayanagi et al., 1992). In another instance, the nucleoid seems to be in a compacted shape when examined by sedimentation and electron microscopy, and the creation of swarmer cells of *Caulobacter crescentus* is linked to the termination of DNA replication. Certain eukaryotes also lack condensation of their chromosomes during the cell cycle. For instance, the majority of dinoflagellates and euglenoids have permanently condensed chromosomes (Rizzo, 1991). On the other hand, in the green alga *Nanochlorum eucaryotum* and the fungus *Zygorhynchus moelleri* (Heath, 1982), chromosomes are dispersed throughout nuclear division (Zahn, 1984). In fact, it is challenging to describe chromosomal segregation in multicellular bacteria without mentioning a process akin to mitosis. Chromosome partitioning in prokaryotes differs from that in most eukaryotes because microtubule (MT)-containing fibers have not been observed linked to bacterial chromosomes. The attachment of bacterial chromosomes to a movable membrane, as suggested for chromosomal propulsion in confined extranuclear pleuromitosis (Van Helvoort et al., 1998), may be necessary for chromosomal movement; nevertheless, (Raikov, 1982). Then, one of the six kinds of eukaryotic mitosis could be homologous to the bacterial process. To further explore this theory, a great deal more research is needed, as it's still possible that DNA polymerase activity provides the force that moves bacterial chromosomes, instead of through a post-replicative mechanism as shown in eukaryotes (Gordon et al., 1998).

3.11.4 FEATURES OF EUKARYOTIC CHROMOSOMES

Since chromosomes contain genes, they are involved in the genetic transmission of traits from one generation to the next. In prokaryotes, the chromosomes are free to roam around in the cytoplasm, but in eukaryotes, they are surrounded by a nuclear membrane. Diverse plant and animal species have diverse chromosome counts, sizes, and shapes. The characteristics of chromosomes include self-duplication, segregation, and mutation. Histones, RNA, and DNA make up chromosomes. The primary genetic component of chromosomes is DNA (Figure 3.1).

3.11.4.1 HISTONES, NUCLEOSOMES, AND CHROMOSOMAL COMPACTION

In the majority of eukaryotes, nucleosomes are formed by DNA being wrapped around histones during chromosomal compaction. That being

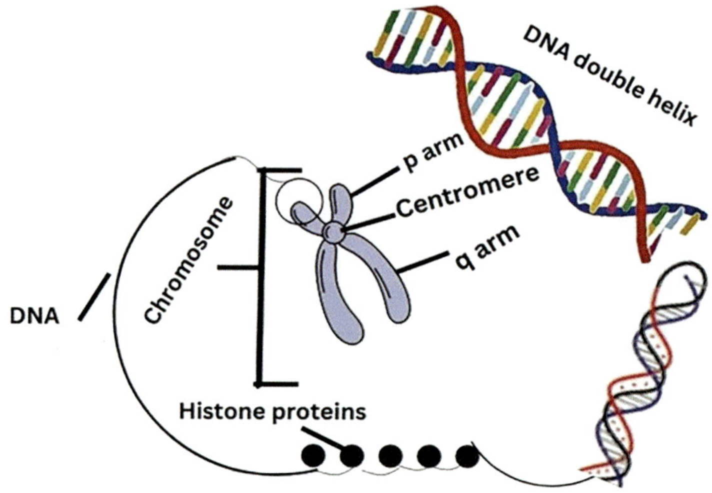

FIGURE 3.1 Structure of eukaryotic chromosome.

said, not all eukaryotes fit this description. Many dinoflagellates have permanently condensed chromosomes, but they lack nucleosomes and histones (Rizzo, 1991). The main inquiry relates to non-nucleosomal organisms and how their DNA is compressed. One intriguing theory is that, as demonstrated *in vitro* with cytoplasmic extracts of *E. coli*, compacting pressures emerge via macromolecular crowding brought on by a high cytoplasmic concentration of non-DNA binding proteins and other large molecules (Zimmerman et al., 1996). In certain organisms, nucleosomes and crowding may bundle the chromosomes at separate points during the life cycle or possibly at the same time. While we have made assumptions about the effects of DNA packing in the crowding type (Drlica et al., 2000).

3.12 CHROMOSOMAL ABERRATIONS

An individual with either monosomy (one missing chromosome from a pair) or trisomy (two or more chromosomes from a pair) will have an abnormally high number of chromosomes, called aneuploidy (Krishna et al., 2019). Both full and partial aneuploidies are possible, where a chromosome is either completely absent or partially inserted. Sexual chromosomes and autosomes can both be aneuploid.

Down syndrome, a developmental disorder caused by an extra copy of chromosome 21, is an example of trisomy in humans. For this reason, the disorder is also known as trisomy 21. Turner syndrome, in which a person is born with only one X sex chromosome, is an example of monosomy in humans (Bersu and Opitz, 1980).

The nuclei contain the thread-like structures called chromosomes. The genetic material is passed down from one generation to the next via the chromosomes. The main structural and functional components of heredity, the genes, are stored in the DNA. Two parallel strands called chromatids, which split apart during cell division, make up the chromosomes. Changes in chromosome number and form are examples of chromosomal abnormalities. Mutations of the chromosomes are another term for them. A common light magnifying instrument can be used to examine chromosomes and reveal several of these mutations. Experts have long utilized these aberrations as a standard for conceptual results in plants since they have been connected to alterations, physical traits, and other features (Preston, 2014). A chromosome's structural aberration can occur spontaneously or be provoked. Such adjustments may result in quantitative changes to genes or in their reorganization. Chromosome breakage and reunification are the root causes of many structural abnormalities. Thus, structural changes are primarily caused by chromosomal breakage. Every broken end has the potential to link to other broken ends, sometimes resulting in new linkage combinations. Depending on how the split ends join and how many breaks there are, a variety of structural changes could occur (Jain et al., 2017).

Various types of chromosomal abnormalities include: three distinct structural changes in chromosomes that have been demonstrated.

- Deletion
- Duplication
- Inversion

A deletion, which causes chromosomal material to be lost, was the first chromosomal aberration that was discovered through genetic evidence. Having more than two copies of a chromosomal segment is referred to as duplication. Inversion is a chromosomal marker for reversible gene order. Sometimes, a section of a chromosomal pair splits off and joins a segment of a non-homologous chromosome, a process known as translocation. Many plants have been reported to exhibit translocations, which have been crucial to the emergence of new species. According to the study's findings, processed industrial effluents like these that are deposited straight into rivers can contain chemicals that harm the chromosomes of exposed species. These pollutants can have detrimental effects on the ecosystem even at trace amounts. Before the effluent is released into the canal, it is recommended that it be completely cleansed to remove any remnants of contaminants that could damage chromosomes (Scientists et al., 2020) (Figure 3.2; Table 3.1).

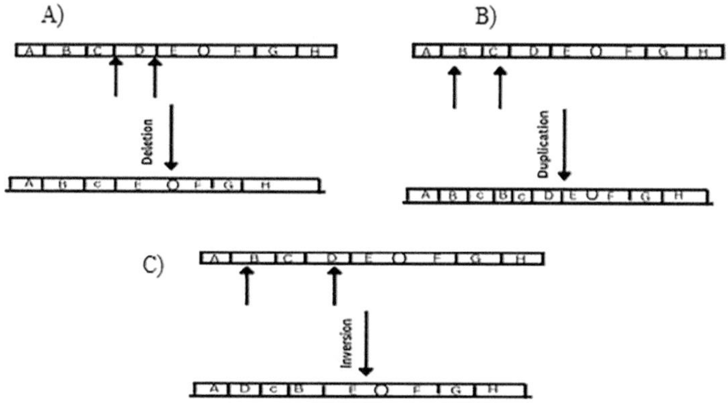

FIGURE 3.2 Illustration of three types of chromosomal structural abnormalities.

TABLE 3.1 Chromosomal Aberrations in an Ecosystem

Factor	Effects	References
Fertilizer	The effects of magnesium sulfate on the cytology of onion roots were investigated. Different magnesium sulfate dosages may cause several chromosomal abnormalities. The current study's findings indicate that magnesium sulfate is cytotoxic and clastogenic.	Bhatta and Sakya (2009)
Hospital wasteful materials	Using the *Allium cepa* test, the genotoxicity of hospital wastewater was evaluated. During the investigation, chromosomal aberrations, bridges, and micronuclei at telophase were discovered; these findings may suggest an environmental contamination.	Bagatini et al. (2009)
MgO presence in plants	Using the *A. cepa* bioassay, the cytotoxicity of MgO nanoparticles and microparticles was evaluated in the plant system.	Mangalampalli et al. (2018)
Herbicides	Toxic to vegetation, especially the tips of roots, is maleic hydrazide. This toxin speeds up evolution by accelerating the pace of change in physical traits and the production of varying aspects to pick increased attributes.	Jabee et al. (2008)

TABLE 3.1 *(Continued)*

Factor	Effects	References
Food Preservatives	Chromosome abnormalities caused by potassium metabisulfite are more common when the concentration of the compound increases. Comparing the amounts to the untreated control, there have been noticeable variances in the amounts. The most common anomalies include gaps, breaks, and repeated breaks.	Panneerselvam (2007)
Curcumin	The results of the investigation demonstrated that curcumin can prevent chromosomal abnormalities in *A. cepa* root tissues caused by sodium azide. Furthermore, curcumin supplementation decreases the mitotic index in all groups, indicating considerable cytotoxicity, while the precise mechanism of action is still unknown.	Ragunathan and Panneerselvam (2007)
Air pollution	Reduced mitotic cell division and the discovery of a genotoxic material	Glasencnik et al. (2004)
Dinocap	Stickiness, laggard/s, multipolarity, c-mitosis, polyploidy fragment/s and micronuclei	Celik et al. (2005)
Fumonisins	Chromosome abnormalities, genetic damage, and sister chromatid exchanges	Lerda et al. (2005)
Lechates from solid waste	Micronuclei, chromosomal abnormalities, and suppression of the mitotic index	Chandra et al. (2005)
Plantago lanceolate	reduce the mitotic index and create stickiness, cracks, and bridges	Celik et al. (2007)
Maleic hydrazide	events mutagenic to decrease and cause chromosomal translocation	Sondhi et al. (2008)
Aluminum	DNA damage, oxidative stress and cellular demise	Achary et al. (2007)
Nano-silver	mitotic index drop, Stickiness, c-metaphase, bridge/s, laggard/s, and micronuclei.	Babu et al. (2008)
Avenoxan	stickiness, aberrant cell growth, bridge/s, laggard/s	Kaymark et al. (2005)
Acetaminophen	roots did not grow in large quantities, and the mitotic index decreased.	Rathore et al. (2005)
Cs and Sr	Onion germination rates are declining; anomalies such as stickiness and vagran.	Bolle et al. (2004)
Cypermeth-rin and fenvalerate	Chromosome and mitotic abnormality; suppression of the mitotic index.	Chauhan et al. (1999)
Phosphine gas	Reduced root length and seed viability, increased incidence of aberrated cells.	Younis et al. (1989)

TABLE 3.1 *(Continued)*

Factor	Effects	References
Sodium benzoate, boric acid, citric acid, potassium citrate and sodium citrate	stickiness, Anaphase bridge/s, break/s, lagging, c-mitosis, micronuclei, and unequal distribution all contribute to mitotic division reduction.	Turkoglu (2007)

3.12.1 SCREENING OF CHROMOSOME ABERRATIONS

For many years, the *Allium cepa* bioassay has been used to evaluate toxicants and the harm they do to the environment and human health (Mangalampalli et al., 2018). The *Allium cepa* test is a useful tool for toxicological screening and monitoring the cytotoxic effects of environmental pollutants. The test has been widely employed to investigate the carcinogenic potential of a number of pesticides, demonstrating that these substances have the ability to cause chromosomal abnormalities in *A. cepa* root stems. Fruits and vegetables that have pesticide residues may be harmful to human health. Chemical fertilizers are known to have harmful and cancerous effects on experimental animals, as well as adverse impacts on human health (Feretti et al., 2007).

KEYWORDS

- **chromosomes**
- **deoxyribonucleic acid**
- **chromosomal theory of heredity**
- ***Allium cepa*** **bioassay**
- **cytogenetics**
- **nucleosome model**
- **environmental toxicology**

REFERENCES

Achary, V. M. M., Jena, S., Panda, K. K., & Panda, B. B. (2007). Aluminum-induced oxidative stress and DNA damage in root cells of *Allium cepa* L. *Ecotoxicology and Environmental Safety, 70*, 300–310.

Adam, M., Robert, F., Larochelle, M., & Gaudreau, L. (2001). H2A.Z is required for global chromatin integrity and for recruitment of RNA polymerase II under specific conditions. *Molecular and Cellular Biology, 21*, 6270–6279.

Ahmad, S. F., & Martins, C. (2019). The modern view of B chromosomes under the impact of high-scale omics analyses. *Cells, 8*(2), 156.

Åkerlund, T., Nordström, K., & Bernander, R. (1995). Analysis of cell size and DNA content in exponentially growing and stationary-phase batch cultures of *Escherichia coli*. *Journal of Bacteriology, 177*, 6791–6797.

Alberts, B., Johnson, A., Lewis, J., Raff, M., Roberts, K., & Walter, P. (2002). The shape and structure of proteins. In *Molecular Biology of the Cell* (4th ed.). Garland Science.

Babu, K., Deepa, M. A., Shankar, S. G., & Rai, S. (2008). Effect of nano-silver on cell division and mitotic chromosomes: A prefatory siren. *The Internet Journal of Nanotechnology, 2*(2).

Bagatini, M. D., Vasconcelos, T. G., Laughinghouse, H. D., Martins, A. F., & Tedesco, S. B. (2009). Biomonitoring hospital effluents by the *Allium cepa* L. test. *Bulletin of Environmental Contamination and Toxicology, 82*(5), 590–592.

Bailey, S. M., & Murnane, J. P. (2006). Telomeres, chromosome instability, and cancer. *Nucleic Acids Research, 34*(8), 2408–2417.

Battaglia, B. (1952). Ricerche sulla spermatogenesi atipica dei Gasteropodi Prosobranchi: II. Le cellule nutrici nella spermatogenesi di *Littorina neritoides* L. (*Gasteropodo Prosobranco*). *Italian Journal of Zoology, 19*(4–6), 195–202.

Battaglia, E. (1955). Chromosome morphology and terminology. *Caryologia, 8*(1), 179–187.

Bernard, P., Maure, J. F., Partridge, J. F., Genier, S., Javerzat, J. P., & Allshire, R. C. (2001). Requirement of heterochromatin for cohesion at centromeres. *Science, 294*, 2539–2542.

Bersu, E. T., & Opitz, J. M. (1980). Anatomical analysis of the developmental effects of aneuploidy in man: The Down syndrome. *American Journal of Medical Genetics, 5*(4), 399–420.

Bhatta, P., & Sakya, S. R. (2009). Study of mitotic activity and chromosomal behaviour in root meristem of *Allium cepa* L. treated with magnesium sulfate. *Ecoprint: An International Journal of Ecology, 15*, 83–88.

Bhaud, Y., Guillebault, D., Lennon, J. F., Defacque, H., Soyer-Gobillard, M. O., & Moreau, H. (2000). Morphology and behaviour of dinoflagellate chromosomes during the cell cycle and mitosis. *Journal of Cell Science, 113*(7), 1231–1239.

Bolle, P., Mastrangelo, S., Tucci, P., & Evandri, M. G. (2004). Clastogenicity of atrazine assessed with the *Allium cepa* test. *Environmental and Molecular Mutagenesis, 43*, 137–141.

Cairns, J. (1962). A minimum estimate for the length of the DNA of *Escherichia coli* obtained by autoradiography. *Journal of Molecular Biology, 4*, 407–409.

Câmara, A. S., Kubalová, I., & Schubert, V. (2023). Helical chromonema coiling is conserved in eukaryotes. *The Plant Journal, 118*(257).

Casjens, S., Murphy, M., DeLange, M., Sampson, L., van Vugt, R., & Huang, W. M. (1997). Telomeres of the linear chromosomes of Lyme disease spirochaetes: Nucleotide sequence and possible exchange with linear plasmid telomeres. *Molecular Microbiology, 26*, 581–596.

Cechova, M., & Miga, K. H. (2022). Satellite DNAs and human sex chromosome variation. In *Seminars in Cell & Developmental Biology* (Vol. *128*, pp. 15–25). Academic Press.

Celik, M., Yuzbasioglu, D., Unal, F., Arslan, O., & Kasap, R. (2005). Effects of dinocap on the mitosis of *Allium cepa* L. *Cytologia, 70*, 13–22.

Celik, T. A., & Aslanturk, O. S. (2007). Cytotoxic and genotoxic effects of *Lavandula stoechas* aqueous extracts. *Biologia, 62*, 292–296.

Chandra, S., Chauhan, L. K. S., Murthy, R. C., Saxena, P. N., Pande, P. N., & Gupta, S. K. (2005). Comparative biomonitoring of leachates from hazardous solid waste of two industries using *Allium* test. *Science of the Total Environment, 347*, 46–52.

Chauhan, L. K. S., Saxena, P. N., & Gupta, S. K. (1999). Cytogenetic effects of cypermethrin and fenvalerate on the root meristem cells of *Allium cepa*. *Environmental and Experimental Botany, 42*, 181–189.

Chen, H. W., Kuspa, A., Keseler, I. M., & Shimkets, L. (1991). Physical map of the *Myxococcus xanthus* chromosome. *Journal of Bacteriology, 173*, 2109–2115.

Christiansen, G., Pedersen, L. B., Koehler, J. E., Lundemose, A. G., & Birkelund, S. (1993). Interaction between the *Chlamydia trachomatis* histone H1-like protein (Hc1) and DNA. *Journal of Bacteriology, 175*, 1785–1795.

Crosland, M. W. J., & Crozier, R. H. (1986). *Myrmecia pilosula*, an ant with only one pair of chromosomes. *Science, 231*, 1278–1281.

Delange, R. J., & Smith, E. L. (1969). Histone function and evolution as viewed by sequence studies. In *Ciba Foundation Symposium 28–The Structure and Function of Chromatin* (pp. 59–76). John Wiley & Sons.

Doolittle, R. F. (1998). Microbial genomes opened up. *Nature, 392*, 339–342.

Drlica, K., & Bendich, A. J. (2000). Chromosome, bacterial. In J. Lederberg (Ed.), *Encyclopedia of Microbiology* (pp. 162–175). Academic Press.

DuPraw, E. J. (1965). The organization of nuclei and chromosomes in honeybee embryonic cells. *Proceedings of the National Academy of Sciences, 53*(1), 161–168.

Feil, R., & Berger, F. (2007). Convergent evolution of genomic imprinting in plants and mammals. *Trends in Genetics, 23*, 192–199.

Feretti, D., Zerbini, I., Zani, C., Ceretti, E., Moretti, M., & Monarca, S. (2007). *Allium cepa* chromosome aberration and micronucleus tests applied to study genotoxicity of extracts from pesticide-treated vegetables and grapes. *Food Additives and Contaminants, 24*(6), 561–572.

Gasparotto, A. E., Milani, D., Martí, E., Ferretti, A. B. S., Bardella, V. B., Hickmann, F., & Cabral-de-Mello, D. C. (2022). A step forward in the genome characterization of the sugarcane borer, *Diatraea saccharalis*: Karyotype analysis, sex chromosome system and repetitive DNAs through a cytogenomic approach. *Chromosoma, 131*(4), 253–267.

Glasencnik, E., Ribaric-Lasnik, C., Savinek, K., Zalubersek, M., Mueller, M., & Batic, F. (2004). Impact of air pollution on genetic material of shallot (*Allium cepa* L. var.

ascalonicum) exposed at differently polluted sites in Slovenia. *Journal of Atmospheric Chemistry, 49*, 363–376.

Gordon, G. S., & Wright, A. (1998). DNA segregation: Putting chromosomes in their place. *Current Biology, 8*, R925–R927.

Gutbrod, M. J., & Martienssen, R. A. (2020). Conserved chromosomal functions of RNA interference. *Nature Reviews Genetics, 21*(5), 311–331.

Hansen, M. T. (1978). Multiplicity of genome equivalents in the radiation-resistant bacterium *Micrococcus radiodurans*. *Journal of Bacteriology, 134*, 71–75.

Heath, I. B., & Rethoret, K. (1982). Mitosis in the fungus *Zygorhynchus moelleri*: Evidence for stage-specific enhancement of microtubule preservation by freeze substitution. *European Journal of Cell Biology, 28*, 180–189.

Herdman, M., Janvier, M., Rippka, R., & Stanier, R. Y. (1979). Genome size of Cyanobacteria. *Journal of General Microbiology, 111*, 73–85.

Heslop-Harrison, J. S., & Schwarzacher, T. (2011). Organization of the plant genome in chromosomes. *The Plant Journal, 66*(1), 18–33.

Hirano, T. (2000). Chromosome cohesion, condensation, and separation. *Annual Review of Biochemistry, 69*, 115–144.

Jabee, F., Ansari, M. Y. K., & Shahab, D. (2008). Studies on the effect of maleic hydrazide on root tip cells and pollen fertility in *Trigonella foenum-graecum* L. *Turkish Journal of Botany, 32*(5), 337–344.

Jain, A. K., Singh, D., Dubey, K., Maurya, R., & Pandey, A. K. (2017). Chromosomal aberrations. In *Mutagenicity: Assays and Applications* (pp. 69–92).

John, B., & Miklos, G. L. G. (1988). *The Eukaryote Genome in Development and Evolution* (p. 416). Allen and Unwin.

Jumas-Bilak, E., Michaux-Charachon, S., Bourg, G., Ramuz, M., & Allardet Servent, A. (1998). Unconventional genomic organization in the alpha subgroup of the Proteobacteria. *Journal of Bacteriology, 180*, 2749–2755.

Kavenoff, R., Klotz, L. C., & Zimm, B. H. (1974). On the nature of chromosome-sized DNA molecules. In *Cold Spring Harbor Symposia on Quantitative Biology* (Vol. *38*, pp. 1–8). Cold Spring Harbor Laboratory Press.

Kaymark, F., & Muranli, F. D. G. (2005). The cytogenetic effects of avenoxan on *Allium cepa* and its relation with pollen sterility. *Acta Biologica Hungarica, 56*(3, 4), 313–321.

Kim, S. K., & Wang, J. C. (1999). Gene silencing via protein-mediated subcellular localization of DNA. *Proceedings of the National Academy of Sciences, 96*, 8557–8561.

Korber, P., & Becker, P. B. (2010). Nucleosome dynamics and epigenetic stability. *Essays in Biochemistry, 48*, 63–74.

Kosak, S. T., Skok, J. A., Medina, K. L., Riblet, R., Le Beau, M. M., Fisher, A. G., & Singh, H. (2002). Subnuclear compartmentalization of immunoglobulin loci during lymphocyte development. *Science, 296*, 158–162.

Krasikova, A., Fishman, V., & Kulikova, T. (2023). Lampbrush chromosome studies in the post-genomic era. *BioEssays, 45*(5), 2200250.

Krishna Chandran, R., Geetha, N., Sakthivel, K. M., Suresh Kumar, R., Jagathnath Krishna, K. M. N., & Sreedharan, H. (2019). Impact of additional chromosomal aberrations on the disease progression of chronic myelogenous leukemia. *Frontiers in Oncology, 9*, 88.

Lerda, D., Bistoni, M. B., Peralta, N., Ychari, S., Vazquez, M., & Bosio, G. (2005). Fumonisins in foods from Cordoba (Argentina), presence and genotoxicity. *Food and Chemical Toxicology, 43*, 691–698.

Lessie, T. G., Hendrickson, W., Manning, B. D., & Devereux, R. (1996). Genomic complexity and plasticity of *Burkholderia cepacia*. *Federation of European Microbiological Societies Letters, 144*, 117–128.

Lin, J., Qi, R., Aston, C., Jing, J., Anantharaman, T. S., Mishra, B., White, O., Daly, M. J., Minton, K. W., & Venter, J. C. (1999). Whole-genome shotgun optical mapping of *Deinococcus radiodurans*. *Science, 285*, 1558–1562.

Maeshima, K., Ide, S., & Babokhov, M. (2019). Dynamic chromatin organization without the 30-nm fiber. *Current Opinion in Cell Biology, 58*, 95–104.

Malandrin, L., Huber, H., & Bernander, R. (1999). Nucleoid structure and partition in *Methanococcus jannaschii*: An archaeon with multiple copies of the chromosome. *Genetics, 152*, 1315–1323.

Maldonado, R., Jimenez, J., & Casadesus, J. (1994). Changes of ploidy during the *Azotobacter vinelandii* growth cycle. *Journal of Bacteriology, 176*, 3911–3919.

Mandrioli, M., & Manicardi, G. C. (2020). Holocentric chromosomes. *PLOS Genetics, 16*(7), e1008918.

Mangalampalli, B., Dumala, N., & Grover, P. (2018). *Allium cepa* root tip assay in assessment of toxicity of magnesium oxide nanoparticles and microparticles. *Journal of Environmental Sciences (China), 66*, 125–137.

Mateuca, R., Lombaert, N., Aka, P. V., Decordier, I., & Kirsch-Volders, M. (2006). Chromosomal changes: Induction, detection methods and applicability in human biomonitoring. *Biochimie, 88*(11), 1515–1531.

McKinley, K. L., & Cheeseman, I. M. (2016). The molecular basis for centromere identity and function. *Nature Reviews Molecular Cell Biology, 17*(1), 16–29.

McStay, B. (2016). Nucleolar organizer regions: Genomic 'dark matter' requiring illumination. *Genes & Development, 30*(14), 1598–1610.

Mirsky, A. E., & Pollister, A. W. (1946). Chromosin, a desoxyribose nucleoprotein complex of the cell nucleus. *The Journal of General Physiology, 30*(2), 117–148.

Nakamura, T. M., Cooper, J. P., & Cech, T. R. (1998). Two modes of survival of fission yeast without telomerase. *Science, 282*, 493–496.

Olins, A. L., & Olins, D. E. (1974). Spheroid chromatin units (ν bodies). *Science, 183*(4122), 330–332.

Ozkan, H., Tuna, M., & Arumuganathan, K. (2003). Nonadditive changes in genome size during allopolyploidization in the wheat (*Aegilops-Triticum*) group. *Journal of Heredity, 94*(3), 260–264.

Padeganeh, A., De Rop, V., & Maddox, P. S. (2013). Nucleosomal composition at the centromere: A numbers game. *Chromosome Research, 21*, 27–36.

Palani, K. L., & Panneerselvam, N. (2007). Cytogenetic studies of food preservative in *Allium cepa* root meristem cells. *Medicine and Biology, 14*(2), 60–63.

Palmer, C. G., & Funderburk, S. (2004). Secondary constrictions in human chromosomes. *Cytogenetic and Genome Research, 4*(4, 5), 261–276.

Peters, J. M., & Nishiyama, T. (2012). Sister chromatid cohesion. *Cold Spring Harbor Perspectives in Biology, 4*(11), a011130.

Peyretaillade, E., Biderre, C., Peyret, P., Duffieux, F., Metenier, G., Gouy, M., Michot, B., & Vivares, C. P. (1998). Microsporidian *Encephalitozoon cuniculi*, a unicellular eukaryote with an unusual chromosomal dispersion of ribosomal genes and LSU rRNA reduced to the universal core. *Nucleic Acids Research, 26*, 3513–3520.

Preston, R. J. (2014). Chromosome aberrations. In *Encyclopedia of Toxicology* (3rd ed., Vol. 1). Elsevier.

Ragunathan, I., & Panneerselvam, N. (2007). Antimutagenic potential of curcumin on chromosomal aberrations in *Allium cepa*. *Journal of Zhejiang University: Science B, 8*(7), 470–475.

Raikov, I. B. (1982). *The Protozoan Nucleus, Morphology, and Evolution. Cell Biology Monographs*. Springer-Verlag.

Rathore, H. S., & Chowbey, P. (2005). Prevention of acetaminophen-induced mitodepression with myrobalan in *Allium cepa* model. *Iranian Journal of Pharmacology and Therapeutics, 4*, 100–104.

Reid, R. A., & Leech, R. M. (1980). The nucleus. In *Biochemistry and Structure of Cell Organelles* (pp. 11–38).

Rine, J. (1999). On the mechanism of silencing in *Escherichia coli*. *Proceedings of the National Academy of Sciences of the United States of America, 96*, 8309–8311.

Rizzo, P. J. (1991). The enigma of the dinoflagellate chromosome. *The Journal of Protozoology, 38*, 246–252.

Rodionov, O., Lobocka, M., & Yarmolinsky, M. (1999). Silencing of genes flanking the P1 plasmid centromere. *Science, 283*, 546–549.

Roy, J. K., & Lakhotia, S. C. (2021). Study of polytene chromosomes of *Drosophila melanogaster*. In *Experiments with Drosophila for Biology Courses* (pp. 175–183).

Scientists, Y., National, F., Editorial, D., & Prof, B. (2020). In *9th YSF Symposium Proceedings 2020*.

Setlow, P. (1995). Mechanisms for the prevention of damage to DNA in spores of *Bacillus* species. *Annual Review of Microbiology, 49*, 29–54.

Shvedunova, M., & Akhtar, A. (2022). Modulation of cellular processes by histone and non-histone protein acetylation. *Nature Reviews Molecular Cell Biology, 23*(5), 329–349.

Sondhi, N., Bhardwaj, R., Kaur, S., Kumar, N., & Singh, B. (2008). Isolation of 24-epibrassinolide from leaves of *Aegle marmelos* and evaluation of its antigenotoxicity employing *Allium cepa* chromosomal aberration assay. *Plant Growth Regulation, 54*, 217–224.

Sorokin, H. (1929). Idiograms, nucleoli, and satellites of certain Ranunculaceae. *American Journal of Botany, 16*, 407–420.

Sutton, W. S. (1902). On the morphology of the chromosome group in *Brachystola magna*. *The Biological Bulletin, 4*(1), 24–39.

Sybenga, J. (2000). Somatic segregation in rice. *Genome, 43*(4), 720–722.

Takayanagi, S., Morimura, S., Kusaoke, H., Yokoyama, Y., Kano, K., & Shioda, M. (1992). Chromosomal structure of the halophilic archaebacterium *Halobacterium salinarium*. *Journal of Bacteriology, 174*, 7207–7216.

Therman, E., & Susman, M. (2012). *Human Chromosomes: Structure, Behavior, and Effects*. Springer Science & Business Media.

Turkoglu, S. (2007). Genotoxicity of five food preservatives tested on root tips of *Allium cepa* L. *Mutation Research/Genetic Toxicology and Environmental Mutagenesis, 626*, 4–14.

Van Helvoort, J. M., Huls, P. G., Vischer, N. O., & Woldringh, C. L. (1998). Fused nucleoids resegregate faster than cell elongation in *Escherichia coli* pbpB(Ts) filaments after release from chloramphenicol inhibition. *Microbiology, 144*, 1309–1317.

Winterstein, C., & Ludwig, B. (1998). Genes coding for respiratory complexes map on all three chromosomes of the *Paracoccus denitrificans* genome. *Archives of Microbiology, 169*, 275–281.

Yamaichi, Y., Iida, T., Park, K. S., Yamamoto, K., & Honda, T. (1999). Physical and genetic map of the genome of *Vibrio parahaemolyticus*: Presence of two chromosomes in *Vibrio* species. *Molecular Microbiology, 31*, 1513–1521.

Younis, S. A., Al-Hakkar, Z. S., Al Rawi, F. I., & Hagop, E. G. (1989). Physiological and cytogenetic effects of phosphine gas in *Allium cepa* (L.). *Journal of Stored Products Research, 25*, 25–30.

Zahn, R. K. (1984). A green alga with minimal eukaryotic features: *Nanochlorum eucaryotum*. *Origins of Life, 13*, 289–303.

Zimmerman, S. B., & Murphy, L. D. (1996). Macromolecular crowding and the mandatory condensation of DNA in bacteria. *FEBS Letters, 390*, 245–248.

CHAPTER 4

Mitosis

KHADIJA HAFEEZ, FAIZA AKHTAR, NADIA YASMEEN, and SHAGUFTA PERVEEN*

Department of Botany, Government College University, Faisalabad, Pakistan

*Corresponding author

ABSTRACT

Mitosis is a mechanism by which a cell duplicates its chromosomes before separating them, creating two identical nuclei in preparation for cell division (CD). The molecular apparatus that prokaryotes use for DNA replication and segregation gave rise to mitosis. Interphase is the longest stage of the cell cycle, during which the cell grows and copies its DNA before mitosis begins. During mitosis, chromosomes align, divide, and move into new daughter cells. Metaphase is a phase of CD during mitosis or meiosis. Each chromosome is typically spread out across the cell nucleus. During metaphase, the nucleus dissolves, and the chromosomes condense and align at the center of the dividing cell. In prophase, chromatin—a mix of DNA and protein—condenses. During anaphase, sister chromatids separate and move to opposite ends of the cell. The genetic material of a parent cell divides into two identical daughter cells during the fifth and final phase of mitosis, telophase. After division, mitosis helps maintain a constant number of chromosomes in daughter cells. It also plays a crucial role in the growth and development of multicellular organisms, tissue repair, and maintaining proper cell size.

4.1 MITOSIS

"Eukaryotic cells arrange and partition their chromosomes during mitosis to prepare for cell division (CD). It is carried out by a cellular apparatus

Chromosomal and Genomic Structures and Functions, Volume 1: Chromosomes and Cell Division.
Tariq Ahmad Bhat and Jameel M. Al-Khayri (Eds)

DOI: 10.1201/9781779640253-4

that is mostly made of microtubules (MTs) and the proteins that are connected with them" (McIntosh et al., 2012).

The division of a cell's nucleus into a pair of daughter nuclei, each containing an equal amount of genetic material, initiates the cell cycle phase that is called mitosis. Cytoplasmic division follows the G2 phase, which comes after the nucleus separates.

The process of CD is essential to life, and its disruption can affect tissue homeostasis, organismal development, and health. Understanding mitotic irregularities helps us understand how various disturbances affect the precision of cell division as well as how particular cellular components, chemicals, and enzyme activity influence this process. Mitosis is necessary for both the replacement of injured cells and the multiplication of functioning cells. Aberrations during mitosis might be the root cause of genetic disorders.

4.2 CHARACTERISTICS OF MITOSIS

- Every time a cell divides, two sister cells are produced from the parent cell.
- The cell can also be referred to as equational division of cells since the total number of chromosome in the parent cell and the daughter cell are equal.
- In plants, the process of mitosis encourages the growth of the vegetative parts, like the stems and roots tips.
- During this process, separation and combinations do not take place.
- The mitotic processes have been broken down into various stages.

4.3 ORIGIN OF MITOSIS

The shift from bacterial cells to eukaryotes brought about the biggest ever rush of duplication of genes and originality, as well as the most profound alteration in cell organization since the beginning of life. The fundamental innovations in the coevolutionary theory of eukaryote history, according to Cavalier (2010), were the coordinated beginnings of the endomembrane structure and cytoskeleton, which were later utilized for the co-evolved mitotic apparatus to create the cell nucleus. Many unique genetic developments in eukaryotes were an inevitable result of compartmentalization and unique segregation mechanisms of DNA.

Bacterial (prokaryotes; cells with genetic material separated by external membranes) and eukaryotic (cells with a nucleus that replicate via mitosis) are the only two basic types of cells (Cavalier-Smith, 1998; Stanier, 1970). As ribosomes begin processing messenger RNA (mRNA) throughout transcription, the generally solitary and circularly organized DNA chromosome of bacteria is connected to its external wall of the cytoplasm and sorted by protein motors connected with that membrane. Eukaryotic chromosomes are often numerous and one-dimensional, diffusely attaching to the cell membrane on the surface. Alternatively, throughout the interphase, the phase of the cell cycle while the cell develops, genetic material is expressed, and DNA is duplicated, it is anchored and encircled by a specialized portion of the endomembrane structure (envelope of the nucleus, or NE).

4.3.1 HISTORY OF EUKARYOTE CHROMOSOME

Eukaryotic chromosomes, on the other hand, are compacted during cell division, preventing transcription or reproduction, and they link via their centromeres to the microtubules about the spindle of mitosis, which transports them to their divided cells. Understanding the co-evolution of approximately 27 parts of cells and how they came to be functionally related to the essentially unique eukaryotic life span (Cavalier-Smith, 1987, 1988, 2002) 850 million years ago, a minimum of two billion years after bacteria morphed (Cavalier-Smith, 2006), is therefore necessary to understand the problem of nuclear origins. Meiosis, syngamy (cell and nuclear fusion), and sex, in addition to mitosis, must have evolved together. This result holds regardless of whether the eukaryote tree exists among unikonts (fungi, animals, and three phyla of protozoa) (Richards and Cavalier-Smith, 2005; Stechmann and Cavalier-Smith, 2003) and bikonts (plants, chromists, and all other eukaryotes) in accordance with current arguments regarding roots reclining in Eozoa (Euglenozoa and excavates), probably within Euglenozoa and excavates sensu stricto (Cavalier-Smith, 2010). These places of the root are true: mitochondria, peroxisomes, cilia, Golgi dictyosomes, and centrioles must have also developed prior to the last common ancestors of all extant eukaryotes (Cavalier-Smith, 2006). The most intricate and widespread example of quantum development throughout life's history is this profound change in cell structure, known as eukaryogenesis (Cavalier-Smith, 2002, 2006; Stanier, 1970). Earth used to be a totally bacterial and viral planet without sexuality. Diatoms, butterflies, coral, whales, kelps, and trees are examples of morphologically complex endoskeleton eukaryotes that emerged later.

Preadaptation, radical mutational innovation, and various sequential selection factors usually play a role in the evolution of complex traits (Cavalier-Smith, 2002, 2006; Simpson, 1953). The genesis of the nucleus is complicated by three issues, according to Cavalier-Smith (1988): the development of the complex of nuclear pores (NPC), which is essential for allowing a channel between the nucleoplasm and cytoplasm; the assembly of endomembranes surrounding chromatin (also known as the DNA-histone complex); and the origin of centromeres and the spindles of mitosis, without which nuclear chromosomes cannot be stably inherited. It is impossible to comprehend the creation of the cell nucleus separately from other crucial developments of the eukaryotic cell, as was first argued 30 years ago (Cavalier-Smith, 1975). Understanding the origins of eukaryotes depends on the intracellular coevolution of several cell components that practically interact or have a significant impact on the selective forces posed on one another (Cavalier-Smith, 1987, 2002).

4.3.2 HISTORY OF PROKARYOTES CHROMOSOME

The prokaryote era of evolution, where sex emerged prior to the nuclear envelope, as well as the predominant selection advantages, were exhibited at that time (Cavalier-Smith, 1975). With our current, far more thorough understanding of cell phylogeny, however, the phylogenetic background has significantly changed (Cavalier-Smith, 2002, 2006; Richards and Cavalier-Smith, 2005; Stechmann and Cavalier-Smith, 2003). Furthermore, advances in molecular cell biology have revealed the way nuclei actually assemble (Cavalier-Smith, 2005; King et al., 2006) and function, allowing us to trace the origins of many significant eukaryotic components, such as NPCs, at the molecular level (Bapteste et al., 2005; Devos et al., 2004, 2006; Mans et al., 2004). On the basis of these realizations, I now put forth the first concrete physical process for the evolution of nuclear envelope design, explicate its major genetic ramifications, and argue that competing theories are insufficient.

4.4 THEORIES OF MITOSIS

Before providing the unique theories, I summarize the two fields to put them in context because the area of eukaryogenesis has become hampered by a plethora of contradicting concepts, some of which may not be compatible with established facts.

1. The evolutionary history about the parts of eukaryotic organisms.
2. The development of the cytoskeleton and system of endomembrane.

As the majority of facts and counterarguments are not new, having been published, I also begin my initial clarification about history with a picture of a nucleus and some fundamental, still frequently disregarded concepts of evolution that are applicable to all such large changes in the body's composition. The mechanism of molecular alterations over significant adaptive switches has become more broadened and complicated than a few molecular evolutionists have realized. This background is lengthy because it is crucial to grasp the right evolutionary context: the cell nucleus did not develop on its own, and explanations of how it came to be make little sense without knowing how the endomembrane network, from which its envelope originates as a specialized component, evolved first. The basic concept in understanding eukaryogenesis is the intracellular coevolution of roughly 100 new features.

4.4.1 CYCLE OF EUKARYOTIC CHROMOSOMES

The eukaryotic chromosomal cycle is composed of the steps "replicate," "condense in preparation for mitosis," "segregate (cell division)," and "decondense." (Losada and Hirano, 2001; Mazia, 1987) is an unnoticed text staple. In eukaryotes, sister-chromatid cohesion (SCC) is an essential chromosomal characteristic that affects a number of distinct chromosome cycle transactions and directly interacts with the cell cycle engine. In the colored portion of the chromosome cycle (Hirano, 2000; Nasmyth et al., 2000), which corresponds to the cell cycle stage that cyclin-dependent kinases (CDKs) are in charge of (Blow and Tanaka, 2005; Stillman, 2005), SCC guarantees that the two sister duplexes remain together. After the process of condensation, to make way for chromosome segregation, the protein connections binding sister chromatids together can entirely break

(Hirano, 2000; Nasmyth et al., 2000). Chromosome condensation-decondensation is used as the readout in one-dimensional representation about CD of eukaryotic cells, which starts out with condensed chromosomes at the start of G1 and quickly decondenses, remaining decondensed in the majority of G1, S, and G2 phases before recondensing at the start of the mitotic phase and remaining condensed during the mitotic phase.

While the readout is a number of chromosomes (the separation status), activity begins with the onset of the mitotic phase, coinciding with sister chromatids' segregation during metaphase as a result of the termination of SCC. This causes the chromosome number to double, which is subsequently followed by their immediate segregation, resulting in the numbers dropping back to half of the maximum at cell division. In relation to the chromosomal cycle of bacteria, the current descriptions of these phases in *Caulobacter crescentus*, *Vibrio cholerae*, *Bacillus subtilis*, *Pseudomonas aeruginosa*, and naturally, *E. coli*, are provided below. In reality, discussions on the topic invariably turn *E. coli*-centric due to the prevalence of *E. coli* research, yet findings within other bacteria confirm a general view of the chromosomal cycle of prokaryotes.

4.4.2 CYCLE OF PROKARYOTIC CHROMOSOMES

The conventional chromosome of bacteria is round and tightly packed into an elliptical nucleus, possibly as an array of rose-like sections (Kavenoff and Bowen, 1976). If translation is inhibited, the nucleoid condenses into a toroid (Zimmerman, 2006). The replication of the chromosome occurs via a solitary bubble of replication, or "unibubble" replication, which is seen in the well-known Cairns theta-replication pattern (Cairns, 1963; Wake, 1973) and is occasionally depicted as a "butterfly" to indicate segregation. However, a single higher-level structure, the molecule of DNA condensed into a nucleoid, is transformed into two such structures rather than a straightforward circular DNA. The replication cycle of the bacterial chromosomal cycle is depicted in detail by the replicating nucleoid when the SCC period that follows replication is also displayed.

Similar models, which first appeared in the replication factory of Dingman (1974), are now widely used to explain prokaryotic nucleoid behavior (Deng et al., 2002; Draper and Gober, 2002; Gordon and Wright, 2000; Holmes and Cozzarelli, 2000; Lbner-Olesen and Kuempel,

1992; Lemon and Grossman, 1998; 2001; Reyes-Lamothe et al., 2008; Sawitzke and Austin, 2001). These diagrams clearly show how a bacterial chromosome cycle differs from that of the eukaryotic chromosome cycle because, in the prokaryotic chromosome, events that occur sequentially in the chromosomes of eukaryotes and influence entire chromosomes seem concurrent at the same time and only affect a small part of the chromosome within a dynamic window. That image of the bacterial chromosome cycle, which packs the majority of the prokaryotic genome during the cycle of the chromosome, emphasizes the small quantity of DNA that is duplicated and separated in addition to the typically compressed nucleoids. To situate the bacterial chromosome cycle events within the known framework of the eukaryotic chromosome cycle or to show the differences between the two cycles, we start from the most universal step: DNA replication.

4.5 FUNCTIONS OF MITOSIS

These are the two key purposes of mitosis:

i. Mitosis aids in an organism's development. Mitosis is a type of asexual reproduction in single-celled organisms.
ii. Mitosis assists in replacing harmed tissues. Mitosis begins in the wounded cells when the healthy cells around them are unable to see the injured cells. Over time, the dividing cells cover the damaged cells by overlapping with them.

4.6 STAGES OF MITOSIS

Most cell growth and division cycles involve an "interphase" phase of synthesis during which enough of every other cellular component is created to meet the needs of two cells. During this time, DNA is replicated. Every daughter cell is completely ready to develop and divide again after mitosis since these factors are changed in "cytokinesis."

The biggest challenge a dividing cell faces is accurately segregating the cell's previously duplicated DNA due to the fact that a genome's DNA makes up a significantly larger portion compared to the circumference of the individual cell. Human DNA, for instance, totals about 2 m, although the majority of cells are barely 1/100,000th that size. The mitotic apparatus is responsible for segregating these lengthy strands with remarkable fidelity.

Any creature needs to complete this task successfully in order to succeed. During replication, each DNA duplex creates "cohesions" that link it to its sister (Peters et al., 2008). Known as "chromatids," these linked DNA duplexes are condensed during mitosis into spheres that are smaller than individual cells.

The outcome is the chromosomes that are compacted, which first appeared in the 19th century and are recognized from iconic CD images. Chromosome condensation happens throughout the "prophase" phase of the cycle of mitosis; it must take place for a CD to be effective. When the spindle forms during prometaphase, the chromosomes are arranged in a twofold-symmetric "metaphase" structure. The effective division of chromosomes during "anaphase" is aided by the twofold symmetry established during metaphase. Two fully functional nuclei may develop in various parts of the cell as metaphase chromatids have sufficiently divided, an event known as "telophase." At the conclusion of the ensuing interphase, cytokinesis then results in the production of two distinct cells that can replicate these occurrences.

The precision with which chromosomal segregation is carried out during mitosis is a crucial characteristic. The cultured cells from mammals are much less precise than yeast budding, which makes just 1 error within around 1 lakh replications (Bakhoum et al., 2014; Hartwell and Smith, 1985). For biological success, accurate chromosomal segregation is clearly crucial, although mitosis is remarkable in a different manner. Nanomachines, or the cytoskeletal fiber subunits and protein molecules that operate as their controllers, connections, and motors, are the mechanical implements of a cell. A typical cellular motor is three times as massive as the majority of chromosomes, and mitotic movement is around 10 times as extensive. The spindle must therefore employ tiny parts to complete a vast and crucial task with exactitude.

Additionally, the spindle is responsible for steering the separation of some components of non-chromosomal cells, such as the Golgi complex and centrosomes, which contain centrioles that can serve as the foundation for developing flagella, and these large organelles within cells with few chloroplasts or mitochondria. Therefore, a spindle is a crucial device that merits our best attempts to comprehend. Every kind of life requires CD, but the discomfort it causes when it fails emphasizes how important it is. Chromosome loss typically results in death in unicellular creatures, which is a very bad negative selection! Incorrect mitosis in multicellular organisms causes aneuploidy, which plays a role in the emergence of cancer. Excessive CD is a definite factor in cancer, while insufficient division can exacerbate anemia. It is understandable why so many scientists have been interested in this topic for so long. The knowledge that has resulted is astounding but undoubtedly incomplete, as the text that follows will demonstrate. Even though this chapter's concentration is on eukaryotic cells, there is still enough variety to brush over many intriguing details. Furthermore, prokaryotic chromosomal segregation is now sufficiently understood to exhibit both instructive similarities to and surprising distinctions from mitosis (Erickson et al., 2010).

4.6.1 INTERPHASE

The mitotic phase of the cell cycle only lasts a short while. Contrast this with the much longer interphase, when the cell prepares to divide. G1 (the first gap), S (synthesis), and G2 (the second gap) are the three interphase phases. By producing proteins and cytoplasmic organelles, the cell develops throughout each of the three phases of interphase. However, only the S phase is when chromosome replication takes place. As a result, a cell begins to grow (G1), continues growing during mitosis, swells even more while preparing for division (G2), divides (M), and then repeats the cycle (S) (Blow and Tanaka, 2005).

4.6.2 PROPHASE

4.6.2.1 PROPHASE IN THE NUCLEUS

When chromosomal condensation becomes clearly evident, prophase has officially begun. The "chromatin" strands, or fibers created when DNA wraps around histone octamers, are impacted by this compaction. Each chromosome becomes shorter and thicker during prophase compaction, separating the DNA strands into distinct, observable units. Transcription stops while compaction takes place, mostly due to the transcriptional factors being moved (Martinez-Balbas et al., 1995). Long-standing discussions on the mechanisms underlying the prophase constriction have been hampered by our incomplete knowledge of chromosomal shape. Chromatin is composed of a charged polymer whose organization changes depending on the environment, which has generated debate. A trustworthy method is electron tomography of frozen-hydrated chromatin, which reveals a network of traversing fiber-connected strands, some of which include the protein family "condensins" that can surround a few fragments of double-stranded DNA and impact its bending (Thandani et al., 2012). Supercoiling and scaffolds were suggested as contributing factors in earlier theories of chromosome structure, but more recent research does not support their existence (Uhlmann, 2014). Given the requirement for this polyelectrolyte to condense to balance charges, post-translational changes of the chromatin, such as phosphorylation and histone acetylation, as well as other acetylated genome proteins, are almost certainly connected to compaction. To understand how chromosomal structure changes during compaction, however, more information is needed than is currently accessible. Numerous processes aid the cell in getting ready to divide as the chromatin condenses. The nucleolus fragments occasionally, but more often than not, it releases a large number of nucleoprotein fragments into the nucleoplasm, which then resemble ribosomes. The DNA that had been the nucleolar organizer was likely compressed as a result of this process. Many eukaryotic cells undergo hyperphosphorylation of nuclear pore complex (NPC) and Lamin fiber components, which line the nuclear envelope's interior surface. This hyperphosphorylation is most likely caused by the serine/threonine-specific cyclin-dependent kinase 1 (CDK1). Lamin-B is still connected to the nuclear envelope membranes, but these fibers are distributed as soluble proteins. In the cells of the majority of mammals and all higher plants, when nucleoplasm and cytoplasm interact during mitosis, the lack of Lamin pathways weakens the envelope, making it more vulnerable to other causes. The spread of the envelope is a planned action. Although it has been observed that the expanding spindle's microtubules (MTs) poke at the envelope, the process will still take place in the presence of medications that prevent MT polymerization. Although the activity of protein kinase C and brief increases in cytosolic Ca^{2+} concentration have been connected to envelope dispersion (Steinhardt and Alderton, 1988), the effects of these changes are still unclear. Integral envelope proteins also disperse as the envelope divides and generates vesicles and cisternae that mix with the endoplasmic reticulum (ER) (Yang et al., 1997).

The nuclear membrane does not disintegrate during mitosis in many tiny eukaryotes; instead, a spindle develops inside the nucleus, resulting in a "closed mitosis." The mitotic spindle proteins must penetrate the nucleoplasm for this method to work, possibly employing nuclear localization cues that are controlled by the cell cycle. *Giardia*, a living being from close to the base of the eukaryotic tree, exhibits closed mitosis, but some more closely related organisms include ciliates, amoebae, and a few fungi. There are several gradations between entirely closed and completely open mitoses, as seen within *Chlamydomonas* green algae or in the syncytial blastoderm of *Drosophila* embryos. With the exception of the polar regions, where centrosomes are found in the cytoplasm, mitotic envelopes are unaltered. These "windows" allow MTs to penetrate the nucleoplasm, where they engage in interactions with the chromosomes (O'Toole et al., 2003). Although some intriguing theories have been put forth, it is still unknown why a certain spindle arrangement occurs in a specific cell type (Sazer et al., 2014). This protein kinase CDK1 regulates many prophase activities, including Lamin dispersion, although other mitotic kinases have also been discovered. The centrosomes' growth in size and activity during mitosis is dependent on polo kinase and its relatives, which are localized there (Lane and Nigg, 1996).

In both fruit flies and nematodes, the prophase development of centrosomes depends on the serine/threonine kinase Aurora A (Giet et al., 2002; Hannak et al., 2001). These enzymes' modifications allow for the buildup of g-tubulin ringed complexes, leading to the initiation of additional MTs connected to the centrosome. On chromosomes, other kinases accumulate, particularly near the "centromere," a genomic major constriction where they participate in mitotic quality control. Current mitosis research has a strong focus on determining the substrates and linkages that comprise each one of these regulatory components.

4.6.2.2 PROPHASE CHANGES IN THE CYTOPLASM

Since ribosomes transit mRNA progressively during prophase, a buildup of polysomes results, which inhibits protein production. Although the exact mechanism behind this alteration is unknown, it has been proposed that it protects mRNA during mitosis and enables a quick restart of protein production after division is complete (Sivan et al., 2007). As their constituents react to elevated protein phosphorylation, MTs and microfilaments both depart from their interphase stability and largely disintegrate (Maller, 1986; Vandre et al., 1986). Even when the fibers of stress go, the cell cortex still has some fibrous actin that can be employed for cytokinesis. Additionally, new assemblies of actin can develop and move in circles around the cell in preparation for the production of contractile rings (Mitsushima et al., 2010). New MTs appear as the MTs of interphase vanish. These organelles start the MTs that become the spindle in species with centrosomes. Because of the phosphorylation of MT-related variables, such as "tog" domain protein molecules (Ma and Poon, 2011), which stimulate Stathmin, a small tubulin-binding protein, and tubulin polymerization while preventing its depolymerization (Gadea and Ruderman, 2006), the subsequent MTs are more unstable compared to their interphase cousins. Following CDK1 phosphorylation, intermediate filaments made of vimentin likewise scatter for mitosis (Chou et al., 1990), while cytokeratins endure and sag over the nucleus as the related MTs vanish. After the nuclear membrane has separated, they create a cage that encloses the nucleoplasm (Zieve et al., 1980).

The prophase change of multicellular plants' cytoskeletons that is important for cytokinesis is the formation of an actin and MTs-containing band that encircles the cell within the plasma membrane. This brief "preprophase band" identifies the cortex of the cell and the cytokinesis site when anaphase comes to a conclusion (Lipka et al., 2014; Zhang et al., 1990). It is still unknown how these markers identify the exact position of cytokinesis at the molecular level.

4.6.2.3 PROPHASE RESTRUCTURING IS IMPORTANT

Cytoskeletal changes during mitosis are significant for four reasons. A noteworthy example of cellular economy is the fact that the interphase cytoskeleton was built using many of the same molecules that make up the spindle and the cytokinetic machinery. Animal cells become rounded as a result of the interphase cytoskeleton vanishing, creating asymmetry that encourages equal distribution among the cell's contents and makes it easier for the equipartition of constituents during cytokinesis. Third, the dissolution of interphase polymers reduces the viscosity of the cytoplasm, permitting medium-sized cytoplasmic components to randomly distribute, such as ribosomes, multienzyme complexes, and small vesicles. At this stage of the cell cycle, cytoplasmic streaming ceases, and the normal mechanism of the plant cell to stimulate quick cytoplasmic mixing is absent. Diffuse movements are therefore essential during plant mitosis.

Diffusion has already accomplished the roughly divided number of cellular parts without the assistance of a particular device like the spindle when cytokinesis splits a cell in two. This process helps maintain the daughter cells' capacity for growth and division. Finally, the rounded cell makes it easier for the spindle to reorient as a result of external cues (Fink et al., 2011), which enables cytokinesis to properly position daughter cells. For multicellular plants, spindle orientation is essential because, after cytokinesis, the cells in these plants are immobile in relation to one another. Some cytomembranes experience analogous prophase dispersal. A HeLa cell's Golgi complex disintegrates into many vesicles and tubules during prophase (Barr, 2004). Thousands of these structures are dispersed very uniformly throughout the cytoplasm during metaphase (Shima et al., 1997). However, Golgi vesicles gather close to the spindle poles in some animal cells. On the other hand, the actin cytoskeleton seems to be engaged in segregating Golgi in yeasts (Barr, 2002). Many cells have a distinct behavior for the ER, moving towards the periphery of the cell and resting just below the surface of the plasma membrane (West et al., 2011). Although the exact process behind this realignment is unknown, it doesn't exhibit diffusion-like characteristics. However, successive cell cleavage achieves the same objective of promoting organelle equipartition. Although spindles frequently initiate development before the nuclear membrane disintegrates, prophase is believed to finish at that point. It initially manifests as more active MTs expanding outside of these distinct structures in cells with centrosomes. Centrosomes in a cell could be close together as MT formation starts, yet they can also be far apart and on distinct sides of a nucleus when MT growth starts, leading to the formation of two discrete asters. The same type of cultivated cell can exhibit both methods, and both result in a proper spindle of metaphase. However, if centrosomes separate early, segregation errors are decreased (Silkworth et al., 2011). The final stage of prophase in a closed mitosis is difficult to pinpoint; spindle formation just begins when chromosomal condensation is adequate. However, unlike other organisms, budding yeasts produce spindles when the chromosomes start to duplicate; G2 displays a bipolar spindle. It is evident that there are various ways to reach the metaphase structure. Since these cells lack centrosomes, the initial phases of spindle production within upper plants are fascinating to study. MTs sheath develops surrounding the African blood lily endosperm's prophase nucleus (De Mey et al., 1982). The center of this sheath "anticipates" the mitotic spindle's axis, which will arise a little later. Sections at the two ends of the sheath that are closer to the nuclear membrane constitute the locations where many MTs originate, proliferate, and ultimately reach the nucleus when the nuclear envelope and sheath scatter. Despite the absence of centrosomes in these cells, the pathway to metaphase appears to be relatively comparable to that of the majority of animal cells.

4.6.3 PROMETAPHASE

4.6.3.1 CHROMOSOMES FORM SPECIALIZATIONS FOR SPINDLE ATTACHMENT

One kinetochore is found on each chromatid, and they are the most crucial components of the chromosome for spindle adhesion. Numerous proteins are used to construct these structures. Approximately 100 distinct polypeptides combine to form many protein complexes in human cells, which then combine on specific chromatin (Cheeseman, 2014). The Ndc80 complex, KNL1, and proteins like CLASP1 and CENP-F, which are located within vertebrates and may have potential analogs in other animals, are only a few of the proteins in fibrous form that bind to MT walls in kinetochores. Motor enzymes are another component of the kinetochore that bind to MTs and produce forces that affect chromosomal location and MT dynamics. Kinetochore motor identities are not retained, but their roles are. Despite the incompleteness of the data, it is likely that every single kinetochore binds a negative end-directed motor. The motor protein is dynein in yeast cells and kinesin-14 in mammalian cells. Additionally, kinetochore molecules bind two or more end-directed motors. Several kinesin-8 proteins, as well as, in the cells of animals, a paired form of kinesin-7 (CENP-E), are present in practically all cells. A kinesin that promotes disassembly is likewise bound by the kinetochore. This includes a number of kinesin-8 proteins found in yeasts, which either directly or indirectly cause MT shortening. Animal cell kinetochores may bind kinesin-13, a non-motile catalyst for MT depolymerization. However, it should be noted that recent research on trypanosome kinetochores has revealed a significant departure from the usual pattern (Akiyoshi and Gull, 2014). Further exceptions than are now predicted may be found when the investigation of mitosis extends to creatures with a larger evolutionary range.

4.6.3.2 SPINDLES ARE WELL-DESIGNED TO ATTACH TO CHROMOSOMES IN A USEFUL MANNER

Microtubules (MTs) of the spindle exhibit unstable dynamics, as reported by Goodson and Jonasson (2016), with disaster rates that are around 10 times higher than interphase rates (Belmont et al., 1990). As a result, rising MTs frequently turn into smaller ones. Following that, they have two options: get rescued or resume growth, or they may shrink to nothing, allowing another MT to begin by releasing its γ-tubulin ring complex. MTs probe the area where chromosomes are located in both this manner and via diffusive revolutions (Kalinina et al., 2013), reiterating their potential to come across a kinetochore. When they do, kinetochore binding stabilizes them and removes them from the moving pool. An excellent model of chromosome attachment to the spindle was originally suggested by Kirschner and Mitchison (1986) in the "search and capture" scenario. The spindle's capacity to ensure that MTs connect to all of the kinetochores on time may be explained mathematically quite successfully (Magidson et al., 2015). As soon as the nuclear membrane opens, the MTs initiated by the centrosome spread throughout the nuclear space and separate in an open mitosis. If they come from centrosomes that have already been split apart, they invade the mass of chromosomes from each side, exposing MTs to the chromosomes in a manner that is approximately twice symmetric. The total amount of MTs that take part in this assault is frequently more than the number that the poles of the spindle themselves can initiate in plants as well as animals. According to Kamasaki et al. (2013) and Petry et al. (2013), "augmin," a complex that links to the walls of the currently present MTs and initiates a new polymer, causes an increase in the number of MTs on the chromosomes, increasing the likelihood that they will attach to the spindle. The spindle expands as the centrosomes split in the cells whose centrosomes hadn't already split apart before envelope dispersal. Then, chromosomes adhere to every surface of this expanding bipolar array, frequently adhering to MT walls. Significant rearranging is then needed because their final location will occur at the MT plus end (see below).

Since closed mitoses frequently start chromosomal attachment while the poles still live close to one another, they too need to have mechanisms to restructure their initial attachments and achieve metaphase architecture. Because chromosomes of metaphase both have chromatids, each of which has its kinetochore, it is possible for some chromosomes to randomly have their "sister" kinetochores connected when using MTs developed from both spindle poles; in this instance, further rearrangement is not necessary. The likelihood of this event, however, is insufficient to accurately predict mitosis. Sister kinetochores, for instance, will occasionally adhere to MTs emerging from a single pole. Spindles provide remarkably reliable error correction for this and other errors. Numerous in-depth analyses on how spindles fix flawed chromosomal attachments are available (Bakhoum et al., 2014; Nicklas, 1997). The fundamental ideas and the underlying mechanics will be the main topics of this study.

4.6.3.3 THE SPINDLE CAN FREQUENTLY CORRECT IMPROPER ATTACHMENTS WHEN THEY FORM

Although the first binding with the kinetochore is most frequently through an MT wall, kinetochores occasionally associate with an MT end that just so happens to grow right into them (Magidson et al., 2011). Since MT walls contain more tubulins than are found on their ends, this is not surprising. How to organize components so that each kinetochore attaches to the terminals of MTs that originate from just one spindle pole is the issue that a cell must resolve. When this mechanism fails, a "merotelic" connection—a single kinetochore adhering to both spindle poles—occurs, which is detrimental to correct chromosome segregation. There are numerous methods for preventing or reversing such connections. A good place to start when it comes to getting attachments correct is the knowledge that each duplicated chromosome has a kinetochore, as well as that MTs produced by two spindle poles enter the nucleoplasm in opposite directions. The fact that those kinetochores are continuously within the MTs with which they are mobile attached is more significant. Dynein's negative end-directed activity pulls every kinetochore towards the MT that originated at the pole first. However, as demonstrated through efficient work involving the micromanipulation of chromosomes in living cells (Nicklas et al., 1998), this attachment is not durable; it maintains only when under tension. Whenever sister poles are connected to sister kinetochores, the centromere of those chromosomes is pushed in opposing orientations, promoting attachment stability. As a result of their instability and lack of this trait, the majority of alternative spindle attachment configurations dissolve, whereas appropriate attachments last. Proper spindle attachment is mostly accomplished through this method; nonetheless, MTs can also leave the kinetochore region and interact with MTs that initiate from the pole to strengthen adhesion (Kitamura et al., 2010).

There are also other ways to achieve correct chromosomal attachment (Foley and Kapoor, 2012; Khodjakov and Rieder, 2009). The associated kinetochore is covered in MTs, generating MTs that are in growing at such a rate that any MT on the opposing pole will probably be eliminated if kinetochore dynein tugs a kinetochore poleward, putting that chromosome near a spindle pole. In the meantime, the sister kinetochore is bent to face MTs developing from the opposing pole; such mechanical tactics, along with others, undoubtedly aid in mitotic fidelity. However, kinetochores also contain the NDC80 complex, which is one of the kinetochore components that may be phosphorylated by Aurora B, a type of protein kinase (DeLuca et al., 2011). By lowering MT compatibility, this alteration encourages separation. As a result, up until Aurora B is turned off, enzymatic disruption of MT-kinetochore interactions is possible through "softening." The MT end is likely pulled out of the vicinity where Aurora B could interact via its substrates and sever the connections, distant from an opposing kinetochore, which is how tension at the kinetochore accomplishes this purpose (Dewar et al., 2004; Lampson and Cheeseman, 2010).

4.6.3.4 THE SPINDLE ASSEMBLY CHECKPOINT (SAC) ACTS TO EXTEND PROMETAPHASE BY CORRECTING INADEQUATELY ATTACHED KINETOCHORE POSITIONS

To govern the localization of extra kinetochore components, kinetochores have additional kinase activities, such as MPS1 (London and Biggins, 2014). These activities are likely in accordance with either the MT attachment or the forces generated in the spindles (Joglekar and Aravamudhan, 2016). According to McIntosh (1991), Roberts et al. (1994), and Hardwick and Murray (1995), these kinases are a crucial component of a novel "checkpoint" that delays the beginning until all chromosomes are properly connected to the spindle; anaphase won't occur. The metabolic cascade that starts a polyubiquitin ligase at detached kinetochores and inhibits the APC/C, or anaphase-promoting complex, is the basis for this checkpoint (Hoyt, 2001). This cascade is cut off when the final chromosome is properly linked, most likely as a result of MT binding as well as tension at the kinetochore. Now that crucial proteins have been polyubiquitinated, the proteasome can degrade them, which in turn triggers a separate protease known as "separase." As will be discussed more below, this enzyme weakens the bonds that hold sister kinetochores together, allowing anaphase to begin. In order to accomplish correct spindle attachment, kinetochores are carefully engineered, and there are several ways to do so—possibly even more than we currently realize.

Growth industries have emerged in kinetochore chemistry and physiology. A lot has been discovered about the stages of kinetochores' maturation, assembly, and function as a result of contemporary techniques in genetics, biochemistry, and biophysics. We also comprehend how kinetochore proteins are spatially organized along the axis of the spindle throughout metaphase (Wan et al., 2009), how various fibrous proteins help kinetochores bind to MTs (Cheeseman, 2014), and the way they can move on MTs to increase the likelihood that they will separate correctly (Kapoor et al., 2006). Furthermore, according to Coue et al. (1991), a kinetochore is capable of maintaining a tight grip while the MT depolymerizes on a spindle. It is now fairly clear how this process works in yeasts. Dam1, a protein complex made up of 10 distinct polypeptides, is surrounded by a ring-shaped prolate protein complex in order for it to function (Westermann et al., 2006). This super-complex offers a strong and trustworthy relationship between a spindle and a chromosome MT, which is of significant relevance when a chromosome engages in a single or limited number of small MTs (Volkov et al., 2013). The kinetochores that these rings are linked to are most likely via the complex NDC80 (Tien et al., 2010). Outside of fungi, the Dam1 complex is not present, but similar processes that attach kinetochores to the terminals of moving MTs utilizing alternative proteins are now the focus of significant research in other classes of animals (Schmidt et al., 2012; Volkov et al., 2015).

4.6.3.5 PATHWAYS FOR CREATING A SPINDLE

Not all spindle formations follow the aforementioned paths. Some spindles that go through meiosis are produced largely by MT initiation close to the chromosomes because of a gradient in the quantity of the little GTPase

Ran within its GTP-associated form created by chromosomes (Kalab and Heald, 2008). MTs immediately surround chromosomes in a thicket as a result of this gradient's impact on various MT initiating and interaction factors. Because of the activities of kinesin-5, dynein, or related minus end-directed motors, which help to cluster MT minus ends, the ensuing fibrous assembly sorts into a bipolar array (Heald et al., 1996). The fibrous protein "nuclear mitotic apparatus protein 1" (NUMA1), which interacts with MTs, especially their terminals, and binds dynein (Chakravarty et al., 2004; Elting et al., 2014; Sikirzhytski et al., 2014), is also involved. NUMA1 helps MTs move past neighbors and form the metaphase observation of a bipolar array. Further study has revealed that both of these pathways may function in a single mitotic cell, despite the fact that when this path was initially found, it appeared to be a replacement for the centrosome-mediated activities mentioned (Kalab and Heald, 2008). However, it appears from experiments in which the Ran-GTP gradient-producing cellular machinery is knocked down that this process is not necessary for mitotically dividing cells (Goshima et al., 2007). As a mutant *Drosophila* that is unable to make centrioles matures to adulthood demonstrates, centrosomes appear to be unnecessary as well (Basto et al., 2006). The spindles continue to form in experiments when laser irradiation damages the centrosomes (Khodjakov et al., 2000). Therefore, several spindle-forming processes coexist; it seems that cells employ a variety of spindle-formation techniques to maximize the likelihood that this crucial activity will proceed without problems. Studies of the diversity of life help us understand some of the fundamental aspects of mitosis. When we examine the mitosis of a class of protozoans termed hypermastigote flagellates, for instance, the line between closed and open mitosis is blurred. Barbulanumpha undergoes a closed mitosis because the nuclear envelope doesn't disintegrate. The spindle, however, develops within the cytoplasm and communicates with the chromosomes via the nuclear membrane. Each chromosome is able to experience forces produced in the cytoplasm because each kinetochore produces an MT anchoring site on the cytoplasmic sides of the envelope by attaching to the internal surface of the membrane (Ritter et al., 1978). Therefore, the location of spindle components in the nucleus or cytoplasm is not particularly significant. Sister kinetochores interacting with sister spindle poles is crucial. According to some earlier research on dinoflagellates, certain living things have different spindle designs that could shed light on how mitosis developed when eukaryotes first appeared (Kubai, 1975).

4.6.4 METAPHASE

4.6.4.1 THE CHROMOSOMES TYPICALLY MOVE TO THE SPINDLE EQUATOR FOLLOWING SPINDLE ATTACHMENT TO FORM THE "METAPHASE PLATE"

Chromosome migration towards the spindle's equatorial axis is a typical, though not always present, component of mitosis. This design must reduce the likelihood that one chromosome will be abandoned when the chromosomes separate by putting all of the chromosomes at a similar "starting line" prior to the commencement of their division during anaphase. Although the processes for "congression" to the spindle center are still being studied, various experimental findings shed light on the possible routes. When prometaphase chromosomes are split by exposure to a microbeam of radiation, the region without kinetochores is pushed away from the adjacent spindle pole (Rieder and Salmon, 1994). Animal cells also push their chromosomal arms away from both poles simultaneously, first to the spindle equator, then outwardly based on the spindle axis. This process, known as the "polar ejection force," is likely a factor in how slowly prometaphase chromosomes move toward the spindle equator. The pressure that dynamically unstable MTs experience as they develop into condensed chromatin and run against resistance as they try to elongate can be used as a model for this force. One other concept based on the spindle axis involves the weakly mobile kinesins-4 and -10 that bind to chromatin toward the plus ends of MT walls (Brouhard and Hunt, 2005). The creation of the plate of metaphase is not prevented by RNA interference against one of these motors, but a twofold knockdown does stop it in HeLa cells (Wandke et al., 2012), proving that such motors are an essential part of chromosome movement to the spindle axis.

The kinesin CENP-E with a positive end provides a unique method for congression. This motor, with a kinetochore, can communicate with MTs on the walls whose apexes are not attached to the kinetochore. A motor moves the kinetochore far from the adjacent pole as it travels toward the MT plus ends (Kapoor et al., 2006). Experiments using abnormal meiotic chromosomes with three kinetochores reveal additional possibilities. These chromosomes have two kinetochores that interact with one pole during metaphase as well as with the other. According to Hays et al. (1982) and Hays and Salmon (1990), such chromosomes occupy a position far from the spindle equator where the length of the larger fiber is equal to the sum of the lengths of the two smaller kinetochore-associated fibers. This finding suggests that spindles have the ability to apply KMTs pole-directed forces proportional to their quantity and length, which would aid in securing the metaphase arrangement. Although the exact process generating that force is unknown, Sturgill et al. (2014) speculate that it may rely on a kinesin-12 motor, which is collected in KMT bundles. Similar to spindle formation, it appears likely that several mechanisms are at work to bring about congression. Although various cells may rely more heavily on different mechanisms, all of the methods can work together to produce the same arrangement, providing yet another illustration of how cells are willing to use a variety of mechanisms to accomplish crucial objectives.

4.6.4.2 THE COMPLEX STEADY STATE IS THE METAPHASE SPINDLE

Spindle MTs are impressively active during prometaphase and even after chromosomal congression is finished. The halftime for MTs that are not attached to kinetochores is 30 seconds (Saxton et al., 1984). According to MT fluorescence, the overall amount of polymer is roughly constant; however, the fluctuating stability of MTs requires continual hydrolysis of significant amounts of GTP in order to maintain a constant level of polymer; hence, this is why it is not harmonious. Furthermore, the more dependable spindle MTs, as KMTs, exhibit a “flux” in the direction of the poles of the spindle (Mitchison, 1989). The best way to observe this sluggish motion is through speckle scanning (Waterman-Storer and Salmon, 1997); it moves virtually as quickly as the succeeding anaphase chromosomal motions. Kinesin-13 at the poles of the spindle is required to catalyze the dissolution of MTs at the minus ends and the incorporation of tubulin into kinetochores or other places of plus-end localization (Zhang et al., 2007). To transport interdigitating MTs away from the spindle equator in some cells, this flux also requires the activity of kinesin-5, necessitating further chemical energy to keep the metaphase condition appearing stable (Brust-Mascher et al., 2009). Although creative models have been proposed, the causes of metaphase MT flow are not immediately apparent (Matos et al., 2009). However, because yeast spindles do not undergo flux, mitosis is not required (Maddox et al., 2000).

4.6.4.3 ALTHOUGH THE START OF METAPHASE IS POORLY DEFINED, ONE OF THE MOST SIGNIFICANT TRANSITIONS IN THE ENTIRE CELL CYCLE OCCURS AT ITS CONCLUSION

It is challenging to pinpoint when metaphase begins since the majority of chromosomes oscillate across the spindle axis, with a magnitude that ranges from chromosome to chromosome as well as cell to cell. Therefore, it can be difficult to determine whether congression has actually occurred, while some cells, including epithelial cells of humans grown in culture, exhibit a "sharpening" just before anaphase, when the plate of metaphase starts. Drugs that stop MT production can produce what is known as a "metaphase arrest," but congression does not take place; therefore, the arrest actually refers to a prometaphase arrest. However, it is clear when metaphase ends because that is when sister chromatids start to split. Since it is a process without an end, the change from metaphase into anaphase is important. When both chromosomes are properly attached, the SAC is silenced, which triggers many regulatory proteins, such as securin, to be polyubiquitinated and subsequently subjected to proteolysis, as was already noted, and cyclin B, which inactivates CDK1. The phosphorylation's that were typical of mitotic entry can now be reversed by phosphatases.

Therefore, a significant cell cycle change results from the correct attachment of each chromosome to the spindle. The time between the last chromosomal attachment and the start of anaphase in cultivated mammalian cells is 20 minutes (Rieder et al., 1995).

4.6.5 ANAPHASE

4.6.5.1 THERE ARE TWO MECHANISMS INVOLVED IN THE SEGREGATION OF ANAPHASE CHROMOSOMES, AND EACH OF THEM IS INTRICATE

Each of a metaphase chromosome's two chromatids separates once separase splits the cohesins that had held the sister strands together. Each of the two chromatids within a metaphase chromosome transforms into an anaphase chromosome, making it easy to identify (Uhlmann, 2001). The motions that follow include an elevation in the distance between the sides of the spindle (anaphase B) and a reduction in the separation of each chromosome from the pole it confronts (anaphase A). The typical order of these motions is A followed by B, but certain cells perform both at once while others only perform one of the two. Both polymer dynamics and motor enzymes are engaged in each of the mentioned motions since they all require modifications to MT lengths and movements of MTs with respect to one another. The specific mechanism that contributes most significantly to each process varies depending on the type of cell. MTs have been marked using a variety of techniques, including fluorescent tubulin speckle imaging, photobleaching, and the photoactivation of tubulin fluorescence, to understand the dynamics of spindle MTs that occur after chromosome segregation. All of these methods have improved our knowledge of spindle MT movements. According to Mitchell et al. (1986), tubulin subunit loss at the kinetochores usually causes KMTs to shorten at the start of anaphase A. The only element of anaphase A in yeasts is this; however, many other cells similarly lose components at both ends (Maddox et al., 2002). All of the tubulin degradation in insect spermatocytes takes place at the pole; nevertheless, kinetochores continue to accumulate components throughout anaphase (LaFountain et al., 2004). Occasionally, these losses occur sequentially. We observe variance in mitotic process specifics once more. This variation most likely results from differences within the functions or locations of particular components of the spindle, such as a motor or kinase. Such distinctions, while effectively instructive, shouldn't prevent us from understanding mitosis as a whole. Kinesin-14 or dynein at kinetochores may assist in pushing the chromosomes toward the pole as KMTs depolymerize (Sharp et al., 2000; Yang et al., 2007). Dynein has a tendency to draw the negative end of KMTs poleward because of its binding to NuMA1, which interacts with MT ends in developed animal cells (Elting et al., 2014). However, the discovery that, regarding minus end-directed chromosomal motion, all minus end-directed motors are unnecessary in both fission and budding yeasts has put this motor-dependent situation in a different context (Grishchuk and McIntosh, 2006; Tanaka et al., 2007). In fact, chromosome motion at physiological speeds has been reconfigured in the lab by attaching separated chromosomes to MTs made from a coverslip-associated polymerization initiator, demonstrating that soluble ATP and/or GTP are not required (Coue et al., 1991; Lombillo et al., 1995).

These findings imply that the forces pulling on chromosomes are influenced by MT dynamics. Though both eukaryotes and prokaryotes evolved from a common ancestor, it's possible that a system based on protein dynamics served as the original mechanism for chromosomal segregation. Polymer dynamics also appear to be an explanation for the separation of chromosomes within eubacteria (Moller-Jensen et al., 2003). Since then, that system has evolved in eukaryotes with the addition of motor-dependent pathways, which have likely increased its dependability and given it a selection advantage (McIntosh et al., 2010).

Studies based on the positions of MT endpoints (McIntosh et al., 1979) or employing speckle photography (Maddox et al., 2002) suggest that anaphase B is produced by the relative mobility of the covering non-KMTs that form the interpolar spindle. Fly spindles' sliding is influenced by kinesin-5, which is antagonistic to different motors and MT cross-linkers (Brust-Mascher and Scholey, 2011). However, it's interesting that research in worm and fungal embryos has demonstrated that MT sliding close to the midplane of the spindle also slows the pace of spindle elongation. MTs that sprout from the poles, nonetheless, protrude from the spindle body—known as "astral MTs"—and interact with dynein that is linked to the cell cortex to produce a force for sliding elongation. Spindle elongation is facilitated by these interactions, which pull the spindle poles (Aist et al., 1991; Civelekoglu-Scholey and Scholey, 2007; Fink et al., 2006) toward the cell surface. Thus, either a "rear-wheel" or "front-wheel" run can lead to anaphase B. The amount of anaphase B is frequently so considerable that the MTs implicated clearly elongate. The location of this MT plus ends polymerization, where they overlap near the spindle midplane, has been directly observed using fluorescent tubulin. Further MT sliding is made possible by this polymerization, which prolongs the process of elongation (Saxton and McIntosh, 1987) and ensures that cytokinesis can transfer one set of chromosomes to each daughter since both sets are adequately spaced.

The KMTs vanished during anaphase A when the chromosomes got closer to the poles. Contrarily, the extended interpolar spindles B can persist, and within some cells, they even develop new MTs as a result of chromosome partitioning. The augmin complex concentrates in a specific area towards the minus ends that constitute interpolar MTs in both plants (Nakaoka et al., 2012) and animal (Uehara and Goshima, 2010) cells, particularly as these ends are segregated from both poles by the reforming nuclei. Many new MTs are initiated as a result of the increased γ-tubulin concentration, with the plus ends pointing in the direction of the cell equator.

The number of MTs in the region within the dividing chromosomes rises as a result of both of these factors; the MT plus ends of the peripheral bundle spread to the cell cortex as a result of MT family interdigitating on the spindle midplane. Even though higher cells in plants and animals exhibit similar increases in MT number, the mechanisms behind the next cytokinesis are very dissimilar. According to Pavic-Kaltenbrunner et al. (2007), in animal cells, the MTs bind to a kinesin-6. This kinesin's plus end-directed movement transfers a "central-spindle complex" from the MT tips to the cell's cortex, where it activates a Rho GTPase and helps control cleavage. Many of these MTs are subsequently bundled by the incoming furrow to make the "midbody," which may last for a while in the succeeding interphase. Consequently, proteolysis and full cleavage pinch the midbody off from one cell or the other, occasionally even both, resulting in its loss (Byers and Abramson, 1968). The two spindle poles take on the role of centrosomes as the spindle breaks down, starting new interphase MTs to support the development of the regular interphase cytoskeleton.

4.6.6 TELOPHASE

4.6.6.1 THE DURATION FOR RESTORING A PRODUCTIVE INTERPHASE NUCLEUS IS IN TELOPHASE

The kinase processes that regulate the cell cycle are changing significantly as the chromosomes segregate. The cell is getting ready to enter interphase once more because CDK1 activity is decreasing, which changes the level of activity of the other kinases that have assisted the cell in entering mitosis. Chromosome arms (CA) shrink, possibly in response to reactivated consolidating movement (Renshaw et al., 2010), bringing the two masses of chromatin closer to the spindle poles. Inner nuclear envelope proteins and maybe Lamins bind to the chromosomes, and fragmented chromosomes that are still condensed begin to rebind to the nuclear membrane (Guttinger et al., 2009).

Such membranes fuse when they come into contact with one another, creating a new membrane that may specify a compartment of the nucleus. Eventually, NPCs rearrange into the envelope, enabling the cell to restore the regular interaction between the cytoplasm and nucleoplasm. Chromatin decondensation begins around this time. Transcription can resume as the process continues, and nucleoli can reform. Although they all appear to be separate from cytoskeletal activity, these mechanisms are all regulated by the cell cycle. In order to achieve this, a mitotic exit pathway regulates the interaction within the termination. Eventually, NPCs rearrange into the envelope (Gupta et al., 2013). This aids in maintaining the coordination of nuclear activities with spindle disintegration and the restoration of the interphase cytoplasm in each of the daughter cells.

4.7 CYTOKINESIS

Instead of being a stage of cytokinesis, cytokinesis is a unique process necessary for finishing the mitotic process. When the metaphase plate once existed, a furrow of cleavage (pinch) comprising a contractile ring now forms in animal cells, pinching off the divided nuclei (Glotzer, 2005). Vesicles produced by the apparatus known as the Golgi, which travel along microtubules to the center of the cell, in both plants and animals, are likewise in charge of promoting cytokinesis (Albertson, 2005). Numerous cells undergo independent cytokinesis and mitosis, leading to single cells with numerous nuclei. Although many diverse organisms exhibit the phenomenon, fungi, slime molds, and coenocytic algae are the most common. When fruit fly embryos are developing at specific phases, for example (Lilly and Duronio, 2005), cytokinesis and mitosis can happen separately.

4.8 SIGNIFICANCE OF MITOSIS

Mitosis is a process that involves nuclear division, which generates two daughter nuclei that are genetically identical to their parent nucleus.

The biological process of mitosis is crucial to many other biological processes and has enormous biological significance:

4.8.1 MULTICELLULAR ORGANISMS' GROWTH

The two daughter cells created have the same number of chromosomes as the original cell's and are genetically related to one another (clones). As a result, unicellular zygotes can split by mitosis and develop into multicellular creatures.

Growth can take place throughout an organism's entire body or only in certain areas, like within the meristems (growing points) of plants.

4.8.2 CELL REPLACEMENT AND TISSUE HEALING

By repairing damaged tissues through mitosis and cell division. Since cells are always dying, they must be continuously replaced by cells that share the same genetic makeup.

Cell replacement happens especially quickly in the skin and gut lining of humans, for instance. Some creatures have the ability to regrow body parts. For example, zebrafish can regrow their fins, while axolotls can regrow their legs and tail, among other body parts.

4.8.3 ASEXUAL PROCREATION

The process of asexual reproduction is when a single organism creates additional members of the same species; the children produced have a genetic connection to the parent.

In unicellular creatures like amoebas, CD results in the production of a genetic progeny that is similar to the original.

Through CD, new multicellular organisms are created from the parent and subsequently separate (or "bud off") in different ways.

Numerous species of plants, fungi, and animals exhibit this sort of reproduction.

Runners from strawberries, yeast, and blossoming in hydra are a few examples of this.

i. An equal number each daughter cell contains chromosomes.
ii. It is in charge of a person's personal growth and development.
iii. It keeps the total number of chromosomes in each body cell of an organism constant.
iv. Mitosis is necessary for vegetative plant growth, asexual reproduction, and the repair and rebuilding of damaged tissues.
v. Mitosis keeps the genome pure because it prevents recombination and crossing over takes place.

4.9 CONCLUSION

A cell arranges many of its parts for equipartition at cytokinesis during mitosis. Through its incorporation in the spindle of mitosis, a device used for the structure and separation of both chromosomes and centrosomes, the MT cytoskeleton performs a crucial function. Numerous aspects and nuanced interactions between proteins, site-specific nucleation regulation of the MT cytoskeleton, protein phosphorylation, and proteolysis have been clarified by research on mitosis. The abundance of data has shed light on how MTs and other related proteins function, contributing to the knowledge that has been accumulated regarding mitotic mechanisms. Although there are still many unanswered questions, the general principles of mitosis are now well known. They serve as a tasteful illustration of cytoplasm engineering and as a testament to the talent and tenacity of the scientific community. However, there are still a number of significant issues regarding the mitotic process. Examples include the unique chromosomal condensation pathways and the interactions between MT ends and both kinetochores and spindle poles. We are aware that even while these connections are being loaded by mitotic force generators, during anaphase, tubulin may add or depart from MT plus terminals on kinetochores, as well as MT minus terminals on the poles. The mechanisms governing these tubulin-MT exchanges, the functions of MT dynamics in the formation of mitotic force, and the pathways by which these processes are controlled are still unknown. The spindle system that allows for error correction is yet another unsolved issue. The majority of the time, chromosomes segregate accurately, although this is partly because the spindle has the capacity to repair errors brought about by chance in bipolar chromosomal attachment during prometaphase. Although it is crucial for the well-being of all higher animals and plants, how faults are identified and fixed is still not well understood. The location of the spindle, which is controlled in many cells and aids in defining the placements of daughter cells that form during mitosis, is another unresolved problem. It is still unclear how spindles are correctly oriented and how body coordinates are sensed. Trypanosomes, dinoflagellates, and ancient algae are only a few examples of the wide range of mitosis that has been observed in cells from various phylogenies. These would also be included in a more thorough explanation of mitosis than the one given here. We could definitely learn more about the mitotic process and perhaps even more about its evolution if we had a better understanding of this diversity.

KEYWORDS

- **cell division**
- **chromosome segregation**
- **error correction mechanisms**
- **microtubule cytoskeleton**
- **mitosis**
- **mitotic phases**
- **spindle apparatus**

REFERENCES

Aist, J. R., Bayles, C. J., Tao, W., & Berns, M. W. (1991). Direct experimental evidence for the existence, structural basis, and function of astral forces during anaphase B *in vivo*. *Journal of Cell Science, 100*(2), 279–288.

Akiyoshi, B., & Gull, K. (2014). Discovery of unconventional kinetochores in kinetoplasts. *Cell, 156*, 1247–1258.

Albertson, R., Riggs, B., & Sullivan, W. (2005). The RE, centrosomes, and Rappaport furrows. *Trends in Cell Biology, 15*(2), 92–101.

Bakhoum, S. F., Silkworth, W. T., Nardi, I. K., Nicholson, J. M., Compton, D. A., & Cimini, D. (2014). The mitotic origin of chromosomal instability. *Current Biology, 24*(4), R148–R149.

Bapteste, E., Charlebois, R. L., MacLeod, D., & Brochier, C. (2005). The two tempos of nuclear pore complex evolution: Highly adapting proteins in an ancient frozen structure. *Genome Biology, 6*(10), 1–15.

Barr, F. A. (2002). Inheritance of the endoplasmic reticulum and Golgi apparatus. *Current Opinion in Cell Biology, 14*(4), 496–499.

Barr, F. A. (2004). Golgi inheritance: Shaken but not stirred. *The Journal of Cell Biology, 164*(7), 955.

Basto, R., Lau, J., Vinogradova, T., Gardiol, A., Woods, C. G., Khodjakov, A., & Raff, J. W. (2006). Flies without centrioles. *Cell, 125*(7), 1375–1386.

Belmont, L. D., Hyman, A. A., Sawin, K. E., & Mitchison, T. J. (1990). Real-time visualization of cell cycle-dependent changes in microtubule dynamics in cytoplasmic extracts. *Cell, 62*(3), 579–589.

Blow, J. J., & Tanaka, T. U. (2005). The chromosome cycle: Coordinating replication and segregation—Second in the cycles review series. *EMBO Reports, 6*(11), 1028–1034.

Brouhard, G. J., & Hunt, A. J. (2005). Microtubule movements on the arms of mitotic chromosomes: Polar ejection forces quantified *in vitro*. *Proceedings of the National Academy of Sciences, 102*(39), 13903–13908.

Brust-Mascher, I., & Scholey, J. M. (2011). Mitotic motors and chromosome segregation: The mechanism of anaphase B. *Biochemical Society Transactions, 39*(5), 1149–1153.

Brust-Mascher, I., Sommi, P., Cheerambathur, D. K., & Scholey, J. M. (2009). Kinesin-5–dependent poleward flux and spindle length control in *Drosophila* embryo mitosis. *Molecular Biology of the Cell, 20*(6), 1749–1762.

Byers, B., & Abramson, D. H. (1968). Cytokinesis in HeLa: Post-telophase delay and microtubule-associated motility. *Protoplasma, 66*, 413–435.

Cairns, J. (1963). The chromosome of *Escherichia coli*. In *Cold Spring Harbor Symposia on Quantitative Biology, 28*, 43–46.

Cavalier-Smith, T. (1975). The origin of nuclei and of eukaryotic cells. *Nature, 256*(5517), 463–468.

Cavalier-Smith, T. (1987). The origin of eukaryote and archaebacterial cells. *Annals of the New York Academy of Sciences, 503*(1), 17–54.

Cavalier-Smith, T. (1988). Origin of the cell nucleus. *BioEssays, 9*(2, 3), 72–78.

Cavalier-Smith, T. (1998). A revised six-kingdom system of life. *Biological Reviews, 73*(3), 203–266.

Cavalier-Smith, T. (2002). The neomuran origin of archaebacteria, the negibacterial root of the universal tree, and bacterial mega classification. *International Journal of Systematic and Evolutionary Microbiology, 52*(1), 7–76.

Cavalier-Smith, T. (2002). The phagotrophic origin of eukaryotes and phylogenetic classification of protozoa. *International Journal of Systematic and Evolutionary Microbiology, 52*(2), 297–354.

Cavalier-Smith, T. (2005). Economy, speed, and size matter: Evolutionary forces driving nuclear genome miniaturization and expansion. *Annals of Botany, 95*(1), 147–175.

Cavalier-Smith, T. (2006). Cell evolution and Earth history: Stasis and revolution. *Philosophical Transactions of the Royal Society B: Biological Sciences, 361*(1470), 969–1006.

Cavalier-Smith, T. (2006). Rooting the tree of life by transition analyses. *Biology Direct, 1*(1), 1–83.

Cavalier-Smith, T. (2010). Kingdoms *Protozoa* and *Chromista* and the eozoan root of the eukaryotic tree. *Biology Letters, 6*(3), 342–345.

Cavalier-Smith, T. (2010). Origin of the cell nucleus, mitosis, and sex: Roles of intracellular coevolution. *Biology Direct, 5*, 1–78.

Chakravarty, A., Howard, L., & Compton, D. A. (2004). A mechanistic model for the organization of microtubule asters by motor and non-motor proteins in a mammalian mitotic extract. *Molecular Biology of the Cell, 15*(5), 2116–2132.

Cheeseman, I. M. (2014). The kinetochore. *Cold Spring Harbor Perspectives in Biology, 6*, a015826.

Chou, Y. H., Bischoff, J. R., Beach, D., & Goldman, R. D. (1990). Intermediate filament reorganization during mitosis is mediated by p34cdc2 phosphorylation of vimentin. *Cell, 62*(6), 1063–1071.

Civelekoglu-Scholey, G., & Scholey, J. M. (2007). Mitotic motors: Kinesin-5 takes a brake. *Current Biology, 17*(14), R544–R547.

Coue, M., Lombillo, V. A., & McIntosh, J. R. (1991). Microtubule depolymerization promotes particle and chromosome movement *in vitro*. *The Journal of Cell Biology, 112*(6), 1165–1175.

De Mey, J., Lambert, A. M., Bajer, A. S., Moeremans, M., & De Brabander, M. (1982). Visualization of microtubules in interphase and mitotic plant cells of *Haemanthus* endosperm with the immuno-gold staining method. *Proceedings of the National Academy of Sciences, 79*(6), 1898–1902.

DeLuca, K. F., Lens, S. M., & DeLuca, J. G. (2011). Temporal changes in Hec1 phosphorylation control kinetochore–microtubule attachment stability during mitosis. *Journal of Cell Science, 124*(4), 622–634.

Deng, W., Burland, V., Plunkett III, G., Boutin, A., Mayhew, G. F., Liss, P., & Perry, R. D. (2002). Genome sequence of *Yersinia pestis* KIM. *Journal of Bacteriology, 184*(16), 4601–4611.

Devos, D., Dokudovskaya, S., Alber, F., Williams, R., Chait, B. T., Sali, A., & Rout, M. P. (2004). Components of coated vesicles and nuclear pore complexes share a common molecular architecture. *PLoS Biology, 2*(12), e380.

Devos, D., Dokudovskaya, S., Williams, R., Alber, F., Eswar, N., Chait, B. T., & Sali, A. (2006). Simple fold composition and modular architecture of the nuclear pore complex. *Proceedings of the National Academy of Sciences, 103*(7), 2172–2177.

Dewar, H., Tanaka, K., Nasmyth, K., & Tanaka, T. U. (2004). Tension between two kinetochores suffices for their bi-orientation on the mitotic spindle. *Nature, 428*(6978), 93–97.

Dingman, C. (1974). Bidirectional chromosome replication: Some topological considerations. *Journal of Theoretical Biology, 43*(1), 187–195.

Draper, G. C., & Gober, J. W. (2002). Bacterial chromosome segregation. *Annual Review of Microbiology, 56*(1), 567–597.

Elting, M. W., Hueschen, C. L., Udy, D. B., & Dumont, S. (2014). Force on spindle microtubule minus ends moves chromosomes. *Journal of Cell Biology, 206*(2), 245–256.

Erickson, H. P., Anderson, D. E., & Osawa, M. (2010). FtsZ in bacterial cytokinesis: Cytoskeleton and force generator all in one. *Microbiology and Molecular Biology Reviews, 74*(4), 504–528.

Fink, G., Schuchardt, I., Colombelli, J., Stelzer, E., & Steinberg, G. (2006). Dynein-mediated pulling forces drive rapid mitotic spindle elongation in *Ustilago maydis*. *The EMBO Journal, 25*(20), 4897–4908.

Fink, J., Carpi, N., Betz, T., Bétard, A., Chebah, M., Azioune, A., & Piel, M. (2011). External forces control mitotic spindle positioning. *Nature Cell Biology, 13*(7), 771–778.

Foley, E. A., & Kapoor, T. M. (2013). Microtubule attachment and spindle assembly checkpoint signalling at the kinetochore. *Nature Reviews Molecular Cell Biology, 14*(1), 25–37.

Gadea, B. B., & Ruderman, J. V. (2006). Aurora B is required for mitotic chromatin-induced phosphorylation of Op18/Stathmin. *Proceedings of the National Academy of Sciences, 103*(12), 4493–4498.

Giet, R., McLean, D., Descamps, S., Lee, M. J., Raff, J. W., Prigent, C., & Glover, D. M. (2002). *Drosophila* Aurora A kinase is required to localize D-TACC to centrosomes and to regulate astral microtubules. *The Journal of Cell Biology, 156*(3), 437–451.

Glotzer, M. (2005). The molecular requirements for cytokinesis. *Science, 307*(5716), 1735–1739.

Glotzer, M. (2017). Cytokinesis in metazoa and fungi. *Cold Spring Harbor Perspectives in Biology, 9*(10), a022343.

Goodson, H. V., & Jonasson, E. M. (2018). Microtubules and microtubule-associated proteins. *Cold Spring Harbor Perspectives in Biology, 10*(6), a022608.

Gordon, G. S., & Wright, A. (2000). DNA segregation in bacteria. *Annual Review of Microbiology, 54*(1), 681–708.

Goshima, G., Wollman, R., Goodwin, S. S., Zhang, N., Scholey, J. M., Vale, R. D., & Stuurman, N. (2007). Genes required for mitotic spindle assembly in *Drosophila* S2 cells. *Science, 316*(5823), 417–421.

Grishchuk, E. L., & McIntosh, J. R. (2006). Microtubule depolymerization can drive poleward chromosome motion in fission yeast. *The EMBO Journal, 25*(20), 4888–4896.

Gupta, S., Mana-Capelli, S., McLean, J. R., Chen, C. T., Ray, S., Gould, K. L., & McCollum, D. (2013). Identification of SIN pathway targets reveals mechanisms of crosstalk between NDR kinase pathways. *Current Biology, 23*(4), 333–338.

Güttinger, S., Laurell, E., & Kutay, U. (2009). Orchestrating nuclear envelope disassembly and reassembly during mitosis. *Nature Reviews Molecular Cell Biology, 10*(3), 178–191.

Hannak, E., Kirkham, M., Hyman, A. A., & Oegema, K. (2001). Aurora-A kinase is required for centrosome maturation in *Caenorhabditis elegans*. *The Journal of Cell Biology, 155*(7), 1109–1116.

Hardwick, K. G., & Murray, A. W. (1995). Mad1p, a phosphoprotein component of the spindle assembly checkpoint in budding yeast. *The Journal of Cell Biology, 131*(3), 709–720.

Hartwell, L. H., & Smith, D. (1985). Altered fidelity of mitotic chromosome transmission in cell cycle mutants of *S. cerevisiae*. *Genetics, 110*(3), 381–395.

Hays, T. S., & Salmon, E. D. (1990). Poleward force at the kinetochore in metaphase depends on the number of kinetochore microtubules. *The Journal of Cell Biology, 110*(2), 391–404.

Hays, T. S., Wise, D., & Salmon, E. D. (1982). Traction force on a kinetochore at metaphase acts as a linear function of kinetochore fiber length. *The Journal of Cell Biology, 93*(2), 374–382.

Heald, R., Tournebize, R., Blank, T., Sandaltzopoulos, R., Becker, P., Hyman, A., & Karsenti, E. (1996). Self-organization of microtubules into bipolar spindles around artificial chromosomes in *Xenopus* egg extracts. *Nature, 382*(6590), 420–425.

Hirano, T. (2000). Chromosome cohesion, condensation, and separation. *Annual Review of Biochemistry, 69*(1), 115–144.

Holmes, V. F., & Cozzarelli, N. R. (2000). Closing the ring: Links between SMC proteins and chromosome partitioning, condensation, and supercoiling. *Proceedings of the National Academy of Sciences, 97*(4), 1322–1324.

Joglekar, A. P., & Aravamudhan, P. (2016). How the kinetochore switches off the spindle assembly checkpoint. *Cell Cycle, 15*(1), 7, 8.

Kaláb, P., & Heald, R. (2008). The RanGTP gradient–a GPS for the mitotic spindle. *The Journal of Cell Biology, 121*(10), 1577–1586.

Kalinina, I., Nandi, A., Delivani, P., Chacón, M. R., Klemm, A. H., Ramunno-Johnson, D., & Tolić-Nørrelykke, I. M. (2013). Pivoting of microtubules around the spindle pole accelerates kinetochore capture. *Nature Cell Biology, 15*(1), 82–87.

Kamasaki, T., O'Toole, E., Kita, S., Osumi, M., Usukura, J., McIntosh, J. R., & Goshima, G. (2013). Augmin-dependent microtubule nucleation at microtubule walls in the spindle. *Journal of Cell Biology, 202*(1), 25–33.

Kapoor, T. M., Lampson, M. A., Hergert, P., Cameron, L., Cimini, D., Salmon, E. D., & Khodjakov, A. (2006). Chromosomes can congress to the metaphase plate before biorientation. *Science, 311*(5759), 388–391.

Kavenoff, R., & Bowen, B. C. (1976). Electron microscopy of membrane-free folded chromosomes from *Escherichia coli*. *Chromosoma, 59*(2), 89–101.

Khodjakov, A., & Rieder, C. L. (2009). Mitosis: Too much of a good thing (can be bad). *Current Biology, 19*(22), 1032–1034.

Khodjakov, A., Cole, R. W., Oakley, B. R., & Rieder, C. L. (2000). Centrosome-independent mitotic spindle formation in vertebrates. *Current Biology, 10*(2), 59–67.

King, M. C., Lusk, C., & Blobel, G. (2006). Karyopherin-mediated import of integral inner nuclear membrane proteins. *Nature, 442*(7106), 1003–1007.

Kirschner, M., & Mitchison, T. (1986). Beyond self-assembly: From microtubules to morphogenesis. *Cell, 45*(3), 329–342.

Kitamura, E., Tanaka, K., Komoto, S., Kitamura, Y., Antony, C., & Tanaka, T. U. (2010). Kinetochores generate microtubules with distal plus ends: Their roles and limited lifetime in mitosis. *Developmental Cell, 18*(2), 248–259.

König, P., Braunfeld, M. B., Sedat, J. W., & Agard, D. A. (2007). The three-dimensional structure of *in vitro* reconstituted *Xenopus laevis* chromosomes by EM tomography. *Chromosoma, 116*(4), 349–372.

Kubai, D. F. (1976). The evolution of the mitotic spindle. *International Review of Cytology, 43*, 167–227.

LaFountain, J. R., Jr., Cohan, C. S., Siegel, A. J., & LaFountain, D. J. (2004). Direct visualization of microtubule flux during metaphase and anaphase in crane-fly spermatocytes. *Molecular Biology of the Cell, 15*(12), 5724–5732.

Lampson, M. A., & Cheeseman, I. M. (2011). Sensing centromere tension: Aurora B and the regulation of kinetochore function. *Trends in Cell Biology, 21*(3), 133–140.

Lane, H. A., & Nigg, E. A. (1996). Antibody microinjection reveals an essential role for human polo-like kinase 1 (Plk1) in the functional maturation of mitotic centrosomes. *The Journal of Cell Biology, 135*(6), 1701–1713.

Lemon, K. P., & Grossman, A. D. (1998). Localization of bacterial DNA polymerase: Evidence for a factory model of replication. *Science, 282*(5393), 1516–1519.

Lemon, K. P., & Grossman, A. D. (2001). The extrusion-capture model for chromosome partitioning in bacteria. *Genes & Development, 15*(16), 2031–2041.

Lilly, M. A., & Duronio, R. J. (2005). New insights into cell cycle control from the *Drosophila* endocycle. *Oncogene, 24*(17), 2765–2775.

Lipka, E., Gadeyne, A., Stöckle, D., Zimmermann, S., De Jaeger, G., Ehrhardt, D. W., & Müller, S. (2014). The phragmoplast-orienting kinesin-12 class proteins translate the positional information of the preprophase band to establish the cortical division zone in *Arabidopsis thaliana*. *The Plant Cell, 26*(6), 2617–2632.

Løbner-Olesen, A., & Kuempel, P. L. (1992). Chromosome partitioning in *Escherichia coli*. *Journal of Bacteriology, 174*(24), 7883–7889.

Lombillo, V. A., Nislow, C., Yen, T. J., Gelfand, V. I., & McIntosh, J. R. (1995). Antibodies to the kinesin motor domain and CENP-E inhibit microtubule depolymerization-dependent motion of chromosomes *in vitro*. *The Journal of Cell Biology, 128*(1), 107–115.

London, N., & Biggins, S. (2014). Mad1 kinetochore recruitment by Mps1-mediated phosphorylation of Bub1 signals the spindle checkpoint. *Genes & Development, 28*(2), 140–152.

Losada, A., & Hirano, T. (2001). Shaping the metaphase chromosome: Coordination of cohesion and condensation. *BioEssays, 23*(10), 924–935.

Ma, H. T., & Poon, R. Y. (2011). How protein kinases coordinate mitosis in animal cells. *Biochemical Journal, 435*(1), 17–31.

Maddox, P. S., Bloom, K. S., & Salmon, E. D. (2000). The polarity and dynamics of microtubule assembly in the budding yeast *Saccharomyces cerevisiae*. *Nature Cell Biology, 2*(1), 36–41.

Maddox, P., Desai, A., Oegema, K., Mitchison, T. J., & Salmon, E. D. (2002). Poleward microtubule flux is a major component of spindle dynamics and anaphase A in mitotic *Drosophila* embryos. *Current Biology, 12*(19), 1670–1674.

Magidson, V., O'Connell, C. B., Lončarek, J., Paul, R., Mogilner, A., & Khodjakov, A. (2011). The spatial arrangement of chromosomes during prometaphase facilitates spindle assembly. *Cell, 146*(4), 555–567.

Magidson, V., Paul, R., Yang, N., Ault, J. G., O'Connell, C. B., Tikhonenko, I., & Khodjakov, A. (2015). Adaptive changes in the kinetochore architecture facilitate proper spindle assembly. *Nature Cell Biology, 17*(9), 1134–1144.

Maller, J. L. (1986). Mitogenic signaling and protein phosphorylation in *Xenopus* oocytes. *Journal of Cyclic Nucleotide and Protein Phosphor Research, 11*, 543–555.

Mans, B., Anantharaman, V., Aravind, L., & Koonin, E. V. (2004). Comparative genomics, evolution, and origins of the nuclear envelope and nuclear pore complex. *Cell Cycle, 3*(12), 1625–1650.

Martínez-Balbás, M. A., Dey, A., Rabindran, S. K., Ozato, K., & Wu, C. (1995). Displacement of sequence-specific transcription factors from mitotic chromatin. *Cell, 83*(1), 29–38.

Matos, I., Pereira, A. J., Lince-Faria, M., Cameron, L. A., Salmon, E. D., & Maiato, H. (2009). Synchronizing chromosome segregation by flux-dependent force equalization at kinetochores. *The Journal of Cell Biology, 186*(1), 11–26.

Mazia, D. (1987). The chromosome cycle and the centrosome cycle in the mitotic cycle. *International Review of Cytology, 100*, 49–92.

McIntosh, J. R. (1991). Structural and mechanical control of mitotic progression. *Cold Spring Harbor Symposia on Quantitative Biology, 56*, 613–619.

McIntosh, J. R., McDonald, K. L., Edwards, M. K., & Ross, B. M. (1979). Three-dimensional structure of the central mitotic spindle of *Diatoma vulgare*. *The Journal of Cell Biology, 83*(2), 428–442.

McIntosh, J. R., Molodtsov, M. I., & Ataullakhanov, F. I. (2012). Biophysics of mitosis. *Quarterly Reviews of Biophysics, 45*(2), 147–207.

McIntosh, J. R., Volkov, V., Ataullakhanov, F. I., & Grishchuk, E. L. (2010). Tubulin depolymerization may be an ancient biological motor. *Journal of Cell Science, 123*(20), 3425–3434.

Mitchison, T. J. (1986). Polewards microtubule flux in the mitotic spindle: Evidence from photoactivation of fluorescence. *The Journal of Cell Biology, 109*(2), 637–652.

Mitchison, T., Evans, L., Schulze, E., & Kirschner, M. (1986). Sites of microtubule assembly and disassembly in the mitotic spindle. *Cell, 45*(4), 515–527.

Mitsushima, M., Aoki, K., Ebisuya, M., Matsumura, S., Yamamoto, T., Matsuda, M., & Nishida, E. (2010). Revolving movement of a dynamic cluster of actin filaments during mitosis. *Journal of Cell Biology, 191*(3), 453–462.

Møller-Jensen, J., Borch, J., Dam, M., Jensen, R. B., Roepstorff, P., & Gerdes, K. (2003). Bacterial mitosis: ParM of plasmid R1 moves plasmid DNA by an actin-like insertional polymerization mechanism. *Molecular Cell, 12*(6), 1477–1487.

Nakaoka, Y., Miki, T., Fujioka, R., Uehara, R., Tomioka, A., Obuse, C., & Goshima, G. (2012). An inducible RNA interference system in *Physcomitrella patens* reveals a dominant role of augmin in phragmoplast microtubule generation. *The Plant Cell, 24*(4), 1478–1493.

Nasmyth, K., Peters, J. M., & Uhlmann, F. (2000). Splitting the chromosome: Cutting the ties that bind sister chromatids. *Science, 288*(5470), 1379–1384.

Nicklas, R. B. (1997). How cells get the right chromosomes. *Science, 275*(5300), 632–637.

Nicklas, R. B., Campbell, M. S., Ward, S. C., & Gorbsky, G. J. (1998). Tension-sensitive kinetochore phosphorylation *in vitro*. *The Journal of Cell Biology, 111*(21), 3189–3196.

O'Toole, E. T., Giddings, T. H., McIntosh, J. R., & Dutcher, S. K. (2003). Three-dimensional organization of basal bodies from wild-type and δ-tubulin deletion strains of *Chlamydomonas reinhardtii*. *Molecular Biology of the Cell, 14*(7), 2999–3012.

Pavicic-Kaltenbrunner, V., Mishima, M., & Glotzer, M. (2007). Cooperative assembly of CYK-4/MgcRacGAP and ZEN-4/MKLP1 to form the centralspindlin complex. *Molecular Biology of the Cell, 18*(12), 4992–5003.

Peters, J. M., Tedeschi, A., & Schmitz, J. (2008). The cohesin complex and its roles in chromosome biology. *Genes & Development, 22*(22), 3089–3114.

Petry, S., Groen, A. C., Ishihara, K., Mitchison, T. J., & Vale, R. D. (2013). Branching microtubule nucleation in *Xenopus* egg extracts mediated by augmin and TPX2. *Cell, 152*(4), 768–777.

Renshaw, M. J., Ward, J. J., Kanemaki, M., Natsume, K., Nedelec, F. J., & Tanaka, T. U. (2010). Condensins promote chromosome recoiling during early anaphase to complete sister chromatid separation. *Developmental Cell, 19*(2), 232–244.

Reyes-Lamothe, R., Wang, X., & Sherratt, D. (2008). *Escherichia coli* and its chromosome. *Trends in Microbiology, 16*(5), 238–245.

Richards, T. A., & Cavalier-Smith, T. (2005). Myosin domain evolution and the primary divergence of eukaryotes. *Nature, 436*(7054), 1113–1118.

Rieder, C. L., & Salmon, E. D. (1994). Motile kinetochores and polar ejection forces dictate chromosome position on the vertebrate mitotic spindle. *The Journal of Cell Biology, 124*(3), 223–233.

Rieder, C. L., Cole, R. W., Khodjakov, A., & Sluder, G. (1995). The checkpoint delaying anaphase in response to chromosome monoorientation is mediated by an inhibitory signal produced by unattached kinetochores. *The Journal of Cell Biology, 130*(4), 941–948.

Ritter, H., Jr., Inoué, S., & Kubai, D. (1978). Mitosis in *Barbulanympha*. I. Spindle structure, formation, and kinetochore engagement. *The Journal of Cell Biology, 77*(3), 638–654.

Roberts, B. T., Farr, K. A., & Hoyt, M. A. (1994). The *Saccharomyces cerevisiae* checkpoint gene *BUB1* encodes a novel protein kinase. *Molecular and Cellular Biology, 14*(12), 8282–8291.

Sawitzke, J., & Austin, S. (2001). An analysis of the factory model for chromosome replication and segregation in bacteria. *Molecular Microbiology, 40*(4), 786–794.
Saxton, W. M., & McIntosh, J. R. (1987). Interzone microtubule behavior in late anaphase and telophase spindles. *The Journal of Cell Biology, 105*(2), 875–886.
Saxton, W. M., Stemple, D. L., Leslie, R. J., Salmon, E. D., Zavortink, M., & McIntosh, J. R. (1984). Tubulin dynamics in cultured mammalian cells. *The Journal of Cell Biology, 99*(6), 2175–2186.
Sazer, S., Lynch, M., & Needleman, D. (2014). Deciphering the evolutionary history of open and closed mitosis. *Current Biology, 24*(22), 1099–1103.
Schmidt, J. C., Arthanari, H., Boeszoermenyi, A., Dashkevich, N. M., Wilson-Kubalek, E. M., Monnier, N., & Cheeseman, I. M. (2012). The kinetochore-bound Ska1 complex tracks depolymerizing microtubules and binds to curved protofilaments. *Developmental Cell, 23*(5), 968–980.
Sharp, D. J., Rogers, G. C., & Scholey, J. M. (2000). Cytoplasmic dynein is required for poleward chromosome movement during mitosis in *Drosophila* embryos. *Nature Cell Biology, 2*(12), 922–930.
Shima, D. T., Haldar, K., Pepperkok, R., Watson, R., & Warren, G. (1997). Partitioning of the Golgi apparatus during mitosis in living HeLa cells. *The Journal of Cell Biology, 137*, 1211–1228.
Sikirzhytski, V., Magidson, V., Steinman, J. B., He, J., Le Berre, M., Tikhonenko, I., & Khodjakov, A. (2014). Direct kinetochore–spindle pole connections are not required for chromosome segregation. *Journal of Cell Biology, 206*(2), 231–243.
Silkworth, W. T., Nardi, I. K., Paul, R., Mogilner, A., & Cimini, D. (2011). Timing of centrosome separation is important for accurate chromosome segregation. *Molecular Biology Cell, 23*, 401–411.
Simpson, G. G. (1953). *The Major Features of Evolution*. Columbia University Press.
Sivan, G., Kedersha, N., & Elroy-Stein, O. (2007). Ribosomal slowdown mediates translational arrest during cellular division. *Molecular and Cellular Biology, 27*(19), 6639–6646.
Stanier, R. Y. (1970). Some aspects of the biology of cells and their possible evolutionary significance. *Genetic Microbiology, 20*, 1–38.
Stechmann, A., & Cavalier-Smith, T. (2003). The root of the eukaryote tree pinpointed. *Current Biology, 13*(17), R665–R666.
Steinhardt, R. A., & Alderton, J. (1988). Intracellular free calcium rise triggers nuclear envelope breakdown in the sea urchin embryo. *Nature, 332*(6162), 364–366.
Stillman, B. (2005). Origin recognition and the chromosome cycle. *FEBS Letters, 579*(4), 877–884.
Sturgill, E. G., Das, D. K., Takizawa, Y., Shin, Y., Collier, S. E., Ohi, M. D., & Ohi, R. (2014). Kinesin-12 Kif15 targets kinetochore fibers through an intrinsic two-step mechanism. *Current Biology, 24*(19), 2307–2313.
Tanaka, K., Kitamura, E., Kitamura, Y., & Tanaka, T. U. (2007). Molecular mechanisms of microtubule-dependent kinetochore transport toward spindle poles. *The Journal of Cell Biology, 178*(2), 269–281.
Thadani, R., Uhlmann, F., & Heeger, S. (2012). Condensin, chromatin crossbarring, and chromosome condensation. *Current Biology, 22*(23), 1012–1021.

Tien, J. F., Umbreit, N. T., Gestaut, D. R., Franck, A. D., Cooper, J., Wordeman, L., & Davis, T. N. (2010). Cooperation of the Dam1 and Ndc80 kinetochore complexes enhances microtubule coupling and is regulated by Aurora B. *Journal of Cell Biology, 189*(4), 713–723.

Toro, E., & Shapiro, L. (2010). Bacterial chromosome organization and segregation. *Cold Spring Harbor Perspectives in Biology, 2*(2), a000349.

Uehara, R., & Goshima, G. (2010). Functional central spindle assembly requires de novo microtubule generation in the interchromosomal region during anaphase. *Journal of Cell Biology, 191*(2), 259–267.

Uhlmann, F. (2001). Chromosome cohesion and segregation in mitosis and meiosis. *Current Opinion in Cell Biology, 13*(6), 754–761.

Uhlmann, F. (2014). A silent revolution in chromosome biology. *Nature Reviews Molecular Cell Biology, 15*(7), 431–431.

Vandré, D. D., Davis, F. M., Rao, P. N., & Borisy, G. G. (1986). Distribution of cytoskeletal proteins sharing a conserved phosphorylated epitope. *European Journal of Cell Biology, 41*(1), 72–81.

Volkov, V. A., Grissom, P. M., Arzhanik, V. K., Zaytsev, A. V., Renganathan, K., McClure-Begley, T., & McIntosh, J. R. (2015). Centromere protein F includes two sites that couple efficiently to depolymerizing microtubules. *Journal of Cell Biology, 209*(6), 813–828.

Volkov, V. A., Zaytsev, A. V., Gudimchuk, N., Grissom, P. M., Gintsburg, A. L., Ataullakhanov, F. I., & Grishchuk, E. L. (2013). Long tethers provide high-force coupling of the Dam1 ring to shortening microtubules. *Proceedings of the National Academy of Sciences, 110*(19), 7708–7713.

Wake, R. G. (1973). Circularity of the *Bacillus subtilis* chromosome and further studies on its bidirectional replication. *Journal of Molecular Biology, 77*(4), 569–575.

Wan, X., O'Quinn, R. P., Pierce, H. L., Joglekar, A. P., Gall, W. E., DeLuca, J. G., & Salmon, E. (2009). Protein architecture of the human kinetochore microtubule attachment site. *Cell, 137*(4), 672–684.

Wandke, C., Barisic, M., Sigl, R., Rauch, V., Wolf, F., Amaro, A. C., & Geley, S. (2012). Human chromokinesins promote chromosome congression and spindle microtubule dynamics during mitosis. *Journal of Cell Biology, 198*(5), 847–863.

Wang, X., Llopis, P. M., & Rudner, D. Z. (2013). Organization and segregation of bacterial chromosomes. *Nature Reviews Genetics, 14*(3), 191–203.

Waterman-Storer, C. M., & Salmon, E. D. (1997). Microtubule dynamics: Treadmilling comes around again. *Current Biology, 7*(6), 369–372.

West, M., Zurek, N., Hoenger, A., & Voeltz, G. K. (2011). A 3D analysis of yeast ER structure reveals how ER domains are organized by membrane curvature. *Journal of Cell Biology, 193*(2), 333–346.

Yang, L., Guan, T., & Gerace, L. (1997). Integral membrane proteins of the nuclear envelope are dispersed throughout the endoplasmic reticulum during mitosis. *The Journal of Cell Biology, 137*(6), 1199–1210.

Yang, Z., Tulu, U. S., Wadsworth, P., & Rieder, C. L. (2007). Kinetochore dynein is required for chromosome motion and congression independent of the spindle checkpoint. *Current Biology, 17*(11), 973–980.

Zhang, D., Rogers, G. C., Buster, D. W., & Sharp, D. J. (2007). Three microtubule severing enzymes contribute to the "Pacman-flux" machinery that moves chromosomes. *The Journal of Cell Biology, 177*(2), 231–242.

Zhang, D., Wadsworth, P., & Hepler, P. K. (1990). Microtubule dynamics in living dividing plant cells: Confocal imaging of microinjected fluorescent brain tubulin. *Proceedings of the National Academy of Sciences, 87*(22), 8820–8824.

Zieve, G. W., Heidemann, S. R., & McIntosh, J. R. (1980). Isolation and partial characterization of a cage of filaments that surrounds the mammalian mitotic spindle. *The Journal of Cell Biology, 87*(1), 160–169.

Zimmerman, S. B. (2006). Shape and compaction of *Escherichia coli* nucleoids. *Journal of Structural Biology, 156*(2), 255–261.

CHAPTER 5

Meiosis

NADIA YASMEEN, KHADIJA HAFEEZ, FAIZA AKHTAR, and
SHAGUFTA PERVEEN*

Department of Botany, Government College University, Faisalabad, Pakistan

*Corresponding author

ABSTRACT

In eukaryotic organisms, meiosis is a crucial phase of the sexual life cycle. Given that homologous recombination involves proteins with great similarity across all domains of life and that it occurred long before meiosis, the majority of evidence points to this being the case. Meiosis plays a significant role in the life of plants and animals. With the help of meiosis, organisms continue their cycle and reproduce themselves. Just as mitosis and meiosis proceed in the life of organisms, variation occurs from generation to generation with the help of meiosis; different genetic recombination occur, and crossing over occurs in sister chromatids.

5.1 MEIOSIS

During the process of meiosis, "just one cell splits twice, producing four cells containing half of the entire genome. These are eggs and sperms – our sex cell." In eukaryotes, meiosis is a crucial phase of the sexual life cycle. Numerous model animals have revealed its mechanical details, and most of its fundamental characteristics have undergone numerous, frequently incongruent evolutionary interpretations (Lenormand et al., 2016).

Chromosomal and Genomic Structures and Functions, Volume 1: Chromosomes and Cell Division.
Tariq Ahmad Bhat and Jameel M. Al-Khayri (Eds)

DOI: 10.1201/9781779640253-5

5.2 ORIGIN OF MEIOSIS

One of the most remarkable evolutionary mysteries is how meiosis evolved gradually (Smith and Maynard-Smith, 1978). One of the "major advances" made by eukaryotes before their ultimate radiation over a billion years ago is meiosis (Cavalier, 2002; Knoll et al., 2006). Meiosis is a process that all living eukaryotes share, suggesting that it only happened once during the last common ancestor they shared (Ramesh et al., 2005; Speiger et al., 2015).

It is challenging to pinpoint the selective environment that gave rise to its evolution; even so, some indications can be gained by identifying:

1. In which different cells containing mitotic related process—such as repairing of DNA —were repeated in meiosis.
2. Which specific actions went into putting the entire cellular process together.
3. The reasons why some meiosis variants would have been less effective.

5.2.1 ORIGIN OF PLOIDY REDUCTION

Proto-meiosis, also known as reduction cell division (CD), most likely originated in early asexual single-cell eukaryotes. For this, two scenarios have been proposed. First, there is the theory that endoreplication—the replication of the nuclear genome without following CD—caused diploidy by accident, and that the restoration of haploidy was chosen as a remedy. A variation of this concept refers to the flip between ploidy levels because different ecological conditions may promote either haploidy or greater ploidy levels (Mable et al., 1998; Rescan et al., 2006). Depending on their developmental stage or certain environmental cues, several other contemporary eukaryotes likewise exhibit somatic ploidy increases and decreases (Parfrey et al., 2008).

It's possible that syngamy was preferred because it hides recessive harmful mutations in diploids. One problem with this theory is that it's possible that this masking is insufficient to encourage diploidy in asexual organisms (Otto et al., 1992). According to one version of this story, selfish elements like plasmids and transposons "manipulated" early syngamy in order to facilitate its horizontal dissemination (Michod, 1998).

This idea is supported by the fact that domesticated mobile genetic components have evolved numerous times in yeasts to enable mating-type flipping, which permits syngamy in haploid colonies (Rusche et al., 2010).

5.2.2 ORIGIN OF DIFFERENT HOMOLOGUE PAIRING AND MEIOTIC RECOMBINATION

Homologue coupling at the start of prophase I allows homologues to be correctly segregated, which is necessary for meiosis. At this point, non-homologous centromere interaction is also frequently seen; however, it is unclear what role this coupling has in evolution and function (Obeso et al., 2014).

"Synapsis," or the coupling of homologous centromeres and the development of a protein structure called the complex of synaptonemal proteins (Kleckner, 1996), is a process that strengthens chromosome pairing in many organisms. After that, chiasmata are classified as either non-crossovers (NCOs), which result in gene-conversion events, or cross-overs (henceforth referred to as "COs"), which result in the interchange of large amounts of different chromatid parts (Baudat et al., 2013).

Another theory is that self-material related to genetic elements propagated and used the DNA repair factory to create recombination (Archetti, 2003). Another theory holds that the combination of pairing and recombination evolved first as a means of repairing mutational damage brought on by endosymbiosis or increasing oxidative stress resulting from growing atmospheric oxygen levels.

Furthermore, this scenario is inconsistent with the finding that a significant amount of double-strand breaks (DSBs) are actively formed at the beginning of division (Michod, 1988).

5.2.3 ORIGIN OF TWO-STEP MEIOSIS

Meiosis is unique in that it begins with chromosomal doubling (S phase) prior to the onset of meiosis. The first steps seem unnecessary for ploidy reduction (Hurst, 1983). In theory, ploidy reduction might be accomplished via a more straightforward single-step CD that skips the first phase of DNA replication. Two-step meiosis has been explained by three different theories. According to the first, two-step meiosis provides more protection against specific self-centered genetic elements (SGEs) that sabotage the meiotic progeny in which they do not end up, increasing their transmission frequencies (sometimes referred to as "sister killers," as opposed to "sperm killers") (Haig, 1991).

The interference mechanism is much less effective in a two-step phase since it is unclear if the reductional phase of meiosis I is occurring and II. Such ambiguity is further increased by the unique alterations to meiosis that red algae and microsporidia express (Haig, 1993). The majority of automictic alterations of two-step meiosis (Winker and Schnittger, 2013).

5.3 SECONDARY MODIFICATIONS OF MEIOSIS

Meiosis is notably universal among all eukaryotes. Variations can be seen in many species that could provide information on the evolutionary history and mechanical limitations of meiosis. Three of the changes are covered: meiosis in asexual organisms, meiosis in polyploids, and inverted meiosis.

5.3.1 MEIOSIS AND POLYPLOIDY

When three or more chromosomes have enough similarity, they may align to differing degrees to produce multivalent. In eukaryotic organisms, polyploidy is surprisingly prevalent considering the significant issues it causes with meiosis. In diploids, homologous chromosomes identify themselves and align, resulting in bivalents during prophase I. This can happen when different polyploidy results from allopolyploid hybridization or when all chromosomal sets come from the exact species (autopolyploid).

In multivalent cells, recombination inside occurs, resulting in rearrangements and mis segregation of chromosomes, which can lower fertility and produce children with low fitness levels (Madlung et al., 2005; Wang et al., 2010).

The occurrence of efficient polyploid species and lineages suggests that selection by chance can frequently encourage conversions from multi- to bivalents, which will thereafter segregate as in diploids, in light of these negative impacts. It is yet unknown, nevertheless, how these changes are accomplished at the molecular level. This technique, which could be accomplished by increasing CO interference, appears to be especially significant in autopolyploids. In the autotetraploid *Thaliana arenosa*, a number of potential genes that could influence these alterations have been discovered. Because aneuploid gametes are usually produced, polyploidy with asymmetry of chromosomal sets presents an even bigger challenge (Griffiths et al., 2006). In these species, the removal of univalent in one sex while inclusion in the other solves the problem of asymmetrical segregation during meiosis; this results, for instance, in haploid sperm as well as tetraploid oocytes in pentaploid dog roses.

5.3.2 INVERTED MEIOSIS

Sister chromatids split during meiosis II in a normal meiotic process, while homologous chromosomes separate during meiotic division I. It is unknown why meiosis normally proceeds in this order, but strangely, certain species—such as mites, mealybugs, flowering plants, and true bugs—have meiosis in the opposite order. Melters et al. (2012) suggest that inverted meiosis may be a potential answer to some kinetochore geometry issues in such meiosis. As of right now, it has been demonstrated that every creature displaying this "inverted" meiosis has holocentric chromosomes, which means that kinetochores are distributed across the entire chromosome as opposed to particular centromeres. Alternatively, meiotic division arrangements might just be a "frozen accident," meaning that reversing it is challenging and was determined long ago by mistake (preferably with monocentric chromosomes). However, a question is raised by new research that examines human female meiosis in extraordinary detail (Ottolini et al., 2015). Although this resulted in aneuploidies in about 23% of cases, the meticulous sequencing of oocytes was conducted. Therefore, the reason why a single sequence of the meiosis process is usually universally present is still a mystery.

5.3.3 MEIOSIS MODIFICATIONS AND LOSS OF SEX

Many species have given up on traditional sexual reproduction, procreating asexually through the inhibition or modification of meiosis, resulting in diploid eggs that are capable of developing without the need for fertilization. This raises two related questions: how can meiosis give rise to mitosis and asexual reproduction (also known as "apomixis" or "clonal parthenogenesis" in vertebrates and "mitotic apomixis" in plants)? And why are some alterations considerably more common than others?

Examples of asexual reproduction processes derived from meiosis include the repression of one of the two meiotic divisions ("automixis" or "meiotic apomixis"), merging two of the four results of a single meiosis ("within-tetrad mating" in fungi and "automixis" in animals), and chromosome doubling before meiosis (referred to as "endomitosis" or "pre-meiotic doubling").

5.3.4 MEIOSIS

It might be challenging to evolve a system that guarantees exclusive sister pairing—that is, the total lack of non-sister pairing—even if several lizard species appear to exhibit it. Furthermore, since both sister chromatids might have the same DSBs, repairing DSBs that occur prior to doubling would be challenging in such a system (Archetti, 2010). However, unless meiosis can be totally avoided (as during asexual reproduction, for example), alterations to meiosis would likely lead to the development of secondary asexuality, preserving a significant amount of the cell's communication and machinery.

5.3.4.1 MEIOSIS I IN DAPHNIA

Meiosis I in Daphnia is interrupted in the middle, and a typical meiosis II occurs. According to Hiruta et al. (2010), clonal diversity in Daphnia is therefore meiotically generated. Significantly, this points to a potential sequential path for the genesis of asexuality resembling mitosis. Numerous species experience rare automixis, which is the spontaneous production of unfertilized eggs. As a result, recombination may be repressed, and variants of automixis that preserve variability in centromere regions may be favorably selected, ultimately resulting in meiosis-derived asexuality with genetic ramifications identical to those of mitosis.

5.3.4.2 MEIOSIS IN ARABIDOPSIS

It is extremely difficult for angiosperms to establish seed development in absence of endosperm fertilization.

Thus, the early development of asexuality may often parallel the initial evolution of self-pollination, and certain features (such as low recombination rates) may predispose certain species to a successful transition to asexuality.

5.4 MEIOSIS PUNCTUATES LIFE CYCLES

Meiosis is an important step in both sexual and some asexual life cycles that are descended from sexual ancestors. In multicellular eukaryotes, recombination is a biological and genetic bottleneck that occurs at a critical juncture within the diploid and haploid phases, where it is closely linked to reproduction (unlike in many protists).

5.4.1 MEIOSIS TIMING AND ARREST

Early haploid eukaryotes most likely underwent syngamy or endomitosis before meiosis. These days, haploid or diploid phases might predominate in many different kinds of life cycles exhibited by multicellular eukaryotes. The timing of meiosis may have initially determined how long each phase lasted; for example, a multicellular, longer diploid phase most likely developed by delaying meiosis. When haploid mitosis is inhibited in animals, meiosis is followed right away by syngamy. There has been extensive research on the exact sequence of recombination in the animal germline. Female meiosis often stops twice, whereas male meiosis proceeds continually. These "meiotic arrests" are governed by a number of variables that vary throughout animals and are not fully understood.

Furthermore, particular cells are "designed" at an early stage to eventually undergo meiosis (germline), in contrast to fungi, plants, and some algal cells, where every cell's fate is determined much later on.

5.4.1.1 PROPHASE I

Arrest 1 lasts for years until sexual maturity is reached and occurs in prophase I throughout early development. Since fertilization usually releases this arrest, it is possible that it evolved to lessen the possibility of oocytes prematurely breaking down into a parthenogenetic state or of improper DNA replication occurring before fertilization.

The precise moment of arrest 2 is more variable, ranging from initial metaphase in many invertebrates to metaphase II, which occurs in vertebrate species, and the first phase of G1 after meiosis II has occurred in some echinoderms. It is unclear, therefore, what evolutionary relevance its exact timing in various populations holds.

Three theories have been proposed to explain the first arrest. Initially, during the prophase, if there are extended intervals between arrests 1 and 2, I might allow the sister chromatid exchange to repair inadvertent DSBs.

Secondly, should Arrest 1 take place in the germline while an earlier mitotic division occurs, this could potentially reduce the variation in the quantity of dangerous mutations between gametes within an individual. This could be problematic if it becomes possible to replace or eliminate a few faulty gametes or early embryos during reproduction. Third, since a meiotic oocyte that is not dividing cannot divide again into mitosis after it begins, it might be simpler to stop unchecked growth in these meiotic oocytes. Therefore, it's possible that Arrest 1 evolved to regulate (and reduce) the quantity of potentially unnecessary and harmful divisions of mitotic cells in the female germline.

It is unknown if plants experience comparable meiotic arrests. It appears that plants have no stringent mechanisms at all to stop the CD during the meiosis process. Plants will go through meiosis regardless of such significant abnormalities, as opposed to fungi and mammals, which can stop recombination and the cell cycle if defects in the sequence are not repaired (Wijnker and Schnittger, 2013).

5.4.2 MEIOSIS AND EPIGENETIC RESET

In every generation, meiosis and syngamy signify crucial shifts from haploid to diploid stages. It has been proposed that meiosis's main purpose in eukaryotes is to enable epigenetic resetting.

For example, several epigenetic alterations (the cytosine methylation and chromatin marks) affect metazoan development; these changes are preserved irrevocably throughout life and need to be reset twice a generation (during the phases of $n \rightarrow 2n$ & $2n \rightarrow n$ transitions).

This guarantees proper development, the emergence of parent-specific impressions, and could enable processes that restrict the number of consecutive mitoses that can occur (known as the "Hayflick limit"), thereby preventing the growth of tumors. Epigenetic inheritance over generations can result from certain loci eluding these resets. For example, it happens post-meiotically in male plant gametophytes (as demonstrated by *Arabidopsis*) or pre-meiotically in the sexes in the germ line of animals (as demonstrated by mice) (Heard and Martienssen, 2014).

It is unclear what these time disparities mean in terms of evolution. Meiosis might not be the best moment to reset epigenetic modifications. TE mobilization, which can be particularly dangerous during gamete production because several epigenetic mechanisms inhibit TE activity, is made possible by resetting epigenetic markers. Furthermore, meiosis may be particularly susceptible to TE activities for a number of reasons. These consist of the following:

i. Ectopic recombination among TEs.
ii. Deficient synapsis and repair as a result of the meiotic apparatus rearranging itself in response to TE-induced DSBs.
iii. Interference with synapsis as a result of TE transcriptional activity. Different TE silencing mechanisms might have developed to guarantee appropriate TE monitoring during epigenetic resetting, such as those using short RNAs. For instance, in mammalian male and female germlines, along with other potential reasons for meiotic errors, it's also plausible that strict synapsis checkpoints arose, at least in part, to avoid the creation of faulty gametes as a result of TE activity.

5.4.3 MEIOSIS SYMMETRICAL AND ASYMMETRY

Symmetrical meiosis produces four gametes that are viable, while asymmetric meiosis produces just one gamete, which is ancestral. Male meiosis is symmetrical in isogamous eukaryotes, seed-bearing plants, "homosporous" species (including mosses and many ferns), and mammals.

5.4.3.1 ASYMMETRICAL MEIOSIS

Nevertheless, it has undergone numerous evolutionary changes and is present in various ciliates, seed plants, and mammal female meiosis. Unresolved are the selective scenarios underpinning the emergence of meiotic asymmetry. Certain situations, like ciliates, do not require the four meiotic products because sex is achieved through the cytoplasmic exchange of haploid micronuclei, a process known as conjugation. In other situations, a large, resource-rich oocyte produced by asymmetric meiosis in females may favor the formation of just one cell instead of four.

Nevertheless, by going through fewer meiosis, females might also do this symmetrically. Given that the symmetrical meiotic process may not guarantee an equal distribution of resources among the four meiocytes, is it possible that asymmetry in meiosis allows for better control over the allocation of resources to oocytes? One challenge in this regard is the lack of clarity regarding the reasons behind the superior efficiency of female control over resource allocation within meiocytes derived from different or identical meiosis.

One way to counteract this could be to make meiocytes struggle with each other for resources during meiosis. This would make a symmetrically female meiosis more susceptible to SGEs that slant the distribution of resources in their favor, perhaps even by eradicating other meiotic products. Thus, asymmetrical meiosis may have evolved to prevent such expensive competition inside tetrads (Haig, 2010). Therefore, the topic of how asymmetrical female meiosis evolved is still unanswered.

5.4.4 FAIRNESS OF MEIOSIS

The seeming fairness of meiosis is one of its most remarkable characteristics: every allele has a 50% probability of appearing in any particular gamete according to the work of Mendel's first law of inheritance. On the other hand, a lot of SGEs manipulate the meiotic mechanism to boost their odds beyond 50%. There are two classes of these SGEs.

Killer SGEs are the first type; they destroy cells that aren't related to the element. The aforementioned "sister killers" could theoretically function during meiosis, but the vast majority of lethal SGEs that have been discovered thus far act post-meiotically, for example, by eliminating sibling sperm. Meiotic drivers that take advantage of the asymmetry in female meiosis make up the second class.

By migrating preferentially into the meiotic product that will eventually form the eggs or megaspores, these components achieve transmission exceeding 50%. Such genomic conflicts (between paternally and maternally derived alleles, as well as between parent and offspring) are likely the cause of parental imprints that differ between male and female combinations on multiple genes controlling embryo growth (Haig, 2014).

Self-centered genetic elements (SGEs) that express later and skew resource consumption in their favor are similar to this form of meiotic drive, where alleles migrate, preferring where resources can be found (the egg). Parents choose how much to give their children prior to the "meiotic barrier of ignorance," and children struggle for resources "from behind the curtain."

The female meiotic driver of the monkeyflower *Mimulus guttatus* formation provides some of the strongest evidence in favor of this theory. Physically, this element is linked to huge centromere-specific satellite DNA arrays; yet, conclusive proof for a direct centromere function is absent.

Robertsonian fusion chromosomes, which result from the fusion of two acrocentric chromosomes into one metacentric chromosome, are able to act like meiotic drivers and separate more frequently into the subsequent egg during meiosis I (Chmátal et al., 2014).

5.5 MEIOSIS AND RECOMBINATION

The transmission of genetic information between homologous chromosomes is a widespread aspect of meiosis. Although we have talked about its origins, the continuing existence of recombination is even more controversial (Otto, 2009).

5.5.1 THE NUMBER OF CROSSOVERS PER CHROMOSOME: CONSTRAINED OR NOT?

The average number of COs per bivalent in many species seems to follow relatively limited patterns, exhibiting minimal variation in comparison to the chromosomal sizes, which vary across a number of orders of magnitude. The physiological duration (in megabases, Mb) per chromosome and the genetic sequence length (in centimeters, where 50 cm is equal to one CO per bivalent) within species have a very strong correlation (R2 > 0.95).

Variation in trans-acting components, including the location of RNF212 as well as its protein, which influences the tendency for DSBs to produce extra COs, could account for the high correlations seen within most species. In fact, the discovery of loci influencing CO rate variation suggests that CO rates within and among species may evolve quickly (Hunter et al., 2014).

On the other hand, high CO quantities are typically not troublesome in terms of appropriate disjunction.

5.5.2 CROSSOVER INTERFERENCE

When a CO is present in one location, it can significantly lower the chance that another CO will occur nearby or on the exact same bivalent. Although this "crossover interference" is common, nothing is understood about its mechanism or purpose. Interfering with it may limit the occurrence or variation of COs, which may be significant for ensuring proper disjunction. In a variety of species, there are two different types of COs that have been found: Class I COs, which are extremely susceptible to intrusion, as well as Class II COs, which don't constitute Class I COs but are thought to be important in guaranteeing obligatory COs.

For example, similar to autopolyploidy, higher interference may prevent abnormal multivalent segregation by limiting the number of COs to a single chromosome. An alternative interpretation of this theory holds that interference serves as a mechanism to prevent COs from happening nearby, which could weaken the cohesiveness between homologues or cause slippage and canceling out when they include more than two non-interlocking chromatids, which prevents COs from occurring.

Nevertheless, these explanations are unable to account for long-distance interference. CO interference may also be adaptive by dismantling genetic links, according to another theory. Initially, since they neutralize each other's impact on genetic connections, nearby COs may be avoided. Secondly, there has been conjecture that CO interference could lessen the likelihood of co-adapted genetic complexes (supergenes) disintegrating. The fact that some species do not exhibit CO interference and, more significantly, that some evidence suggests that the level of interference evolves in the course of evolution investigations in *Drosophila* lends some credence to the hypothesis that CO disruption is not an entirely mechanistic constraint (Agarwal et al., 2015).

5.5.3 DIFFERENCES IN RECOMBINATION RATES BETWEEN THE SEXES

Crossover (CO) ratios and localization during male and female recombination differ in many species, and even among closely related species. This is the most extreme situation where these variations could differ in magnitude and orientation. Lack of blending with a single sex—always the heterogametic sex—is known as achiasmy. This could have developed as a result of selection driving sex-to-sex chromosomal recombination suppression or as a means of encouraging it. Numerous theories have been put forth, including those pertaining to mechanical elements like variations in the structure of chromatin. Sexual concurrence, gender variation, and dispersal are examples of globally spread evolutionary processes, as is the sexual variation in the functioning of loci connected to CO rate (e.g., RNF212). There is a strong connection without preventing crossing on the Z or X chromosome. More interesting are the quantitative differences between males and females, known as heterochiasmy, which occur across several taxa but whose mechanical and evolutionary origins are yet unknown.

5.5.4 THE LOCALIZATION OF CROSSOVERS AND RECOMBINATION HOTSPOTS

Biased conversion of genes (BGC) during DSB repair erodes the target sequence motifs, causing these hotspots to self-destruct. This creates a "hotspot paradox": in the outermost portion of BGC, how can hotspots as well as recombination be sustained over the long run? A plausible interpretation is that trans-acting factors, like PRDM9, could change fast enough to persistently "seek" emerging and frequently encountered targets (hotspots), changing to new ones when BGC causes these targets to become rare.

This "Red Queen" model closely resembles the hotspot turnover pattern that has been seen in certain instances and does not require substantial stabilization control on the amount of COs. However, since adequate segregation was probably already guaranteed by the earlier kind of hotspots when it emerged.

Furthermore, it fails to explain why the PRDM9 action is self-destructive, as a trans-acting factor does not have to cause DSBs at the precise location of the target sequence. In actuality, fixed chromosomal features may preserve one copy number per chromosome; therefore, it makes no sense to depend on a target sequence for this reason.

5.6 ORIGIN OF MEIOSIS FROM PROKARYOTIC TO EUKARYOTIC

Regarding the origin of meiosis, there are two competing views. One is that, just as eukaryotes emerged from prokaryotes, meiosis also originated from prokaryotic coupling (bacterial recombination) (Bernstein et al., 2010). The second is that mitosis gave rise to meiosis.

5.6.1 FROM PROKARYOTIC SEX

In prokaryotic sex, the information contained in the DNA of one prokaryote is absorbed by another prokaryotic organism and incorporated into the chromosome of that other prokaryote. Donor genetic material can be incorporated into living prokaryotes by conjugation or transformation. The oldest known form of sexual contact may have been transformation, in which the DNA from one prokaryotic organism gets released into the medium that surrounds it after being taken up by another prokaryotic cell. Meiosis emerged from transformation, according to one explanation for its origin. This theory holds that the evolutionary shift from bacterial to eukaryotic sexual reproduction occurred continuously. Similar to meiosis, transformation is a complicated process that depends on several different gene products. One of the main similarities between eukaryotic and prokaryotic sex is that both involve the joining of DNA from two distinct individuals (parents) in order to align homologous sequences, which is followed by the exchange of gene information (a process known as genetic recombination). The newly generated recombinant chromosome is passed on to offspring. RecA is one of the key enzymes that catalyzes this process in bacteria. It functions similarly to two other enzymes that are involved in eukaryotic meiosis, RAD51 and DMC1 (Villeneuve and Hillers, 2001). For example, it was previously believed that the common intestinal parasite organism *Giardia intestinalis*, a basic eukaryotic protozoan, emerged through a recently diverged eukaryotic lineage that did not have sex.

Nevertheless, further research has demonstrated that *G. intestinalis* possesses a core set of meiotic-related genes in its genome, containing five genes that are unique to meiosis. Furthermore, meiosis gene homologs are involved in a specific, sex-like process that *G. intestinalis* recently underwent. This data, together with other such occurrences, points to the presence of an ancestor that underwent primitive meiosis in the common ancestral lineage of all eukaryotes, which originated from an earlier prokaryote.

5.6.2 FROM EUKARYOTIC MITOSIS

Unlike meiosis, which involves dividing chromosomes into two copies and dividing all of each of the daughter cells, mitosis is the usual procedure for cell division in eukaryotes. According to the mitosis theory, meiosis developed from mitosis (Holliday et al., 2009).

This theory states that recombination and sexual reproduction emerged only after early eukaryotes had evolved and established mitosis. Furthermore, Wilkins and Holliday pointed out that meiosis requires four unique stages that are absent from mitosis.

These are the following:

During the first meiotic division, sister chromatid separation is suppressed, homologous chromosomal pairing occurs, substantial recombination occurs between homologs, and chromosome replication is avoided during the second meiotic division.

Wilkins and Holliday contend that just one new step—homolog synapsis—was specifically started in the process of the development of meiosis from mitosis, despite the appearance of complexity in the emergence of these phases. Two additional novel traits, meanwhile, might have developed later through substantial recombination, or they might have been the result of straightforward alterations.

5.7 COEVOLUTION WITH MEIOSIS

Mitosis and meiosis may have developed simultaneously in the early history of eukaryotes if meiosis resulted from prokaryotic transformation. Meiosis originated through the bacterial sexual process of transformation, whereas mitosis arose from the molecular machinery employed by bacterial cells for the replication of DNA and segregation. Both processes have common molecular components. Meiosis, however, also utilizes the dynamic molecular machinery necessary for DNA segregation and replication.

5.8 FUNCTION

A wealth of data suggests that facultative sexual eukaryotes typically reproduce sexually in adverse environments. For example, the single-celled fungus *Saccharomyces cerevisiae*, which is a budding yeast, reproduces asexually (mitotically) as diploid cells while nutrients are plentiful, but under situations of deprivation, it converts to meiosis (sexual reproduction) (Herskowitz, 1988). *Chlamydomonas reinhardtii* is a unicellular green alga that grows as cells of vegetation in nutrient-rich growth medium. However, gamete fusion, zygote development, and meiosis occur when the nitrogen supply in the medium is depleted. When exposed to H_2O_2 to induce oxidative stress, the breakdown of the yeast *Schizosaccharomyces pombe* significantly increases the percentage of cells that go through meiosis. When the simple multicellular eukaryote *Volvox carteri* experiences heat shock or oxidative stress, it engages in sexual activity. These and other instances imply that meiosis is an adaptation that allows single-celled as well as multicellular eukaryotes to deal with stress.

Prokaryotic sexual behavior also seems to be a stress adaptation. For example, transformation happens close to the final stage of logarithmic growth, that is, when cells reach the end of the logarithmic phase in *Haemophilus influenzae* or when amino acid sequences become inadequate in *Bacillus subtilis*. Transformation in *Streptococcus mutans* as well as other streptococci is linked to the creation of biofilms and high cell density.

Transformation in *Streptococcus pneumoniae* is brought about by the DNA-damaging substance mitomycin C. These and other examples suggest that prokaryotic sex is an adaptation to harsh environments, similar to meiosis in basic eukaryotes. This finding implies that the selective forces preserving bacterial sex and the natural selection forces sustaining meiosis in eukaryotes are comparable. This resemblance points to a continuation of sexual evolution from bacterial cells to eukaryotes rather than a break in it.

However, stress is a broad concept. What particular aspect of stress does meiosis need to overcome? And what particular advantage does meiosis offer that improves resilience to adversity?

5.9 DEOXYRIBONUCLEIC ACID (DNA) REPAIR

Meiosis is essentially a process for repairing damage to DNA, according to one idea. Stresses from the environment frequently produce oxidative stress in cells, which is widely known to damage DNA by releasing reactive oxygen species (ROS), which are oxygen molecules that are reactive. If DNA damage is not repaired, it can lead to a cell's death by preventing the transcription of vital genes or DNA replication. Nevertheless, ROS can also result in double-strand damage, which is a hard-to-fix kind of damage.

The double-strand break is a typical illustration of double-strand damage. In this instance, both strands of the damaged chromosome lose their genetic information (nucleotide sequence), and the only way to obtain the correct information is to find an undamaged chromosome that is homologous to the damaged chromosome. Recombination repair is the method that the cell uses to precisely carry out this kind of repair. It is well known that the recombination repair mechanism that occurs during meiosis between non-sister chromosomes is capable of repairing double-strand breaks and other forms of double-strand damage. Recombination between sister chromosomes, on the other hand, is unable to fix double-strand defects that result from replication that happened earlier.

n every single cell generation in humans, there are typically 50 breaks in the double strand per cell. Meiosis can effectively repair these common defects in the genetic material passed on to germ cells, preventing the loss of human fertility.

Meiosis enhances recombinational repairs between non-sister chromosomes. Recombinational repair is hence the selective benefit of meiosis in both single-celled and multicellular eukaryotes, including humans, according to the notion that meiosis evolved from prokaryotic sex.

One argument against this theory is that prokaryotes already have sufficient repair mechanisms, including recombination-based ones (Lenormand et al., 2016). Prokaryotes indeed possess a recombinational repair-enhanced DNA repair mechanism, and the fact that they can survive in harsh environments shows how effective this system is at repairing DNA damage caused by environmental factors.

Nevertheless, the majority of these processes might be more mutagenic and less precise than meiosis' repair mechanism.

For the most part, they can undergo recombination that encourages a more thorough repair without the need for a second homologous chromosome. Furthermore, meiosis was naturally selected because it is a more accurate repair mechanism for any harm that occurs at any stage of the cell cycle than mitotic repair mechanisms, as it involves a greater amount of homologous recombinational repair than mitotic repair mechanisms do.

However, the sister chromatid involved in mitotic recombination may have experienced a comparable level of stress; as a result, rather than mitigating the damage, this kind of recombination may potentially exacerbate it and reduce fitness.

5.10 PROPHASE I ARREST

All the oocytes required for subsequent ovulations are present in female mammal and bird embryos at birth, and all oocytes are stopped at the prophase I stage of meiosis. For instance, oocytes are present at birth in human fetuses because they develop between 3 and 4 months of gestation. In the oocytes throughout this prophase I delayed stage (dictyate), which can endure for several years, there are four copies of the genome.

According to Mira (1998), the idea behind the halting of oocytes at each of the four genomic copy stages was to give the germline DNA the informational redundancy it needs to heal damage. It is most likely homologous recombinational repair that is employed. Oocytes that have been arrested in prophase are very capable of efficiently repairing DNA damage.

Adaptive DNA repair throughout meiosis appears to be an important quality control mechanism and a crucial fertility determinant in the female germline.

5.11 GENETIC DIVERSITY

Stress is thought to be a warning to the individual that the external environment is becoming unfavorable, which would explain another theory on the function of meiosis. It might be advantageous to have offspring with a different genetic makeup from the parents under this new circumstance. Some of these diverse offspring might have adjusted to the new environment better than their parents did. Meiosis produces genetic variety in the mitotic cell, partly through recombination, which is the transfer of genetic material between chromosome pairs once they align. According to Otto and Gerstein (2006), this is why meiosis has the benefit of promoting genetic variety in offspring, which enables adaptation to unfavorable environmental changes.

5.12 SIGNIFICANCE OF MEIOSIS

There are four different hypothesized biological purposes for meiosis.

1. First, the restoration of genetic abnormalities in germline line cells is significantly aided by recombination during meiosis.
2. Secondly, the widely belief that it creates the genetic diversity necessary for natural selection to operate through recombination and sexual reproduction.
3. Third, it contributes for the preservation of the germ line's immortality by preventing damaged meiocytes from proliferating the germ line through the removal of flawed RNA and protein molecules.
4. Fourth, that it is necessary for the retraining of gametes that result in the fertilized egg, at least in mammals.

An integrated theory is put forth in an effort to connect these disparate roles. In higher animals, there is now substantial evidence that the order of methylation of cytosine in DNA, which is passed down through the action of a maintenance methylase, plays a role in controlling gene activation. According to one theory, the lack of a methyl group that is crucial for function in a promoter or operator region creates a signal that triggers recombination. Then, maintenance methylase activity can replace the missing methyl groups in this region, thanks to the development of hybrid DNA.

Additionally, there is proof that meiotic recombination is mostly limited to neighboring DNA or structural genes. If completely new methylation is not possible, epigenetic abnormalities may result from the absence of methyl groups that methylase has no ability to restore in somatic and germline cells. Because it raises the likelihood that epigenetic abnormalities due to different factors could become homozygous and become irreversible through recombination, inbreeding would be detrimental. Sexual reproduction, as well as outbreeding processes, would be beneficial if some epigenetic flaws were somehow passed down to the next generation. This is because they would allow for their removal during the subsequent meiosis (Holliday, 1984). The pairing, synapsis, recombination, and segregation of homologous chromosomes—the key components of meiosis—are necessary for the development of sperm and eggs and are crucial to sexual reproduction. Novel allelic combinations between parents and children, as well as between individual progeny, are produced during recombination as a result of DNA exchange between homologs (Wang and Copenhaver, 2018). The basis for both speciation and biodiversity is this genetic variety. In the course of commercial breeding of plants and animals, new elite qualities are also developed through the phenotypic variety that arises from genetic variance. Thus, comprehending the molecular processes that drive and govern plant meiosis helps expedite the improvement of crops and offers a theoretical framework for the creation and upkeep of novel agricultural types.

5.13 MEIOSIS IN ANIMALS

Because of intersex and hermaphroditism, distinct sexes are not clearly defined entities. Most people believe that diploid sexes originated from hermaphrodite ancestors more than once (Gorelick, 2003). For whatever reason, there has been a widespread belief that one specific genetic locus determines diploid sex, or whether a person is female or male. Nonetheless, a number of biologists have claimed that there cannot be a methodical technique to differentiate between diploid metazoans that are female and male.

The clitoris and penis in humans have the same sensory function, making them similar. Recognizing that heteromorphic sexual chromosomes only exist in a small number of animals across multiple different taxa, heteromorphic sexual activity chromosomes exist. Organelles originate paternally and biparentally in many different plant and animal species, including doubly uniparental mollusks, and as such, cannot be utilized to discriminate between the sexes. One common method used to differentiate between the sexes is the creation of eggs as opposed to sperm. Thus, in many eukaryotes, it appears that all apparent distinctions between diploid males and females are irrelevant.

The variations among sexes of haploid animals, haploid cells, haploid gametes, or haploid nuclei have been a focus of various investigations. This could be because animal gametes are single-celled and supposedly transitory. Since most metazoan eggs need the fertilization process to carry out egg meiosis, the majority of metazoan cells that produce eggs are diploid. During the process of fertilization, haploid females consequently become ephemeral, if not completely absent. This makes it challenging to separate haploids from diploids in most cases. Assigning sexes to different heterogeneous haploids is also not always possible, at least not across all the eukaryotes, because sexual promiscuity exists in numerous haploid organisms (e.g., gametophytes, which are homosporous in nature, ferns that generate both eggs and sperm). For instance, in humans, the fertilization of an egg cell by a sperm cell results in the haploidization of the egg nucleus only following plasmogamy but before karyogamy. Looking for connections between egg size and number is most useful when examining clades with atypically large sperm.

This explains the substantial *Drosophila* research. Nevertheless, there are enormous gaps in our understanding of anisogamy because, to the best of our knowledge, investigations have not been carried out with any of the other types of species. There are at least 12 metazoan phyla that include amoeboid sperm: *Platyhelminthes*, *Annelida*, the arthropods, the Chordata family, Dicyemida, *Gastrotricha*, *Gnatostomulida*, *Nematomorpha*, and *Acanthocephala*. Additionally, sperm size varies greatly in no fewer than two arthropod lineages: hemipterans and ostracods, which both contain enormous taxa.

Despite being incredibly fruitful, the overwhelming focus on drosophilids has also warped fundamental evolutionary terminology. In most contexts, the prefix “iso-” denotes similarity or equality, while its antonym is “aniso-.” Traditionally, the prefix “iso-” does not apply if there is any difference between two entities. As a result, any discernible variation between sperm and eggs is referred to as anisogamy. Nonetheless, research examining the potential negative association between sperm quantity and size due to resource limitations has been characterized by isogamy (Bjork and Pitnick, 2006).

A casual examination of seed plants reveals that the size and number of sperm cells do not correlate inversely. Although the sperm of sphenogram organisms are much smaller than those of zoogamous gymnosperms, the number of sperm in both taxa is roughly equal because both produce roughly the same amount of pollen, which consistently produces two sperm, with the exception of *Microcycas calocoma,* which can produce up to 12 sperm per pollen tube (Norstog, 1993).

5.13.1 ANISOGAMY

Perhaps arguments based on anisogamy should be centered on functional differences rather than size disparities. A typical response is that haploid males, often sperm, only supply the genetic code that controls the growth of the growing embryo, while haploid females, typically eggs, supply the biological mechanism for the developing embryo. This is an old theory that goes back at least as far as Aristotle, and it still holds true in spite of the evidence mentioned above (Cobb, 2006). According to Aristotle, the female specified the matter that would result in the fetus, and the male's sperm contributed the form, which helped shape and sculpt the developing embryo (Mayhew, 2004).

Anisogamy would offer a fine, universal means of differentiating females from males if it originated in a single group of early eukaryotes. Nonetheless, anisogamy is typically thought to have evolved independently numerous times during the evolution of eukaryotes (Bell, 1978).

Egg competition and sperm may have contributed to the independent multiple evolution of anisogamy, among other mechanisms. It seems illogical to use distinct standards to differentiate between males and females, with a single standard for each distinct anisogamy origin. This would not constitute a crucial distinction between the sexes. To complicate things even more, there is a real continuum between anisogamy and isogamy, which is particularly noticeable in green algae.

5.13.1.1 EGGS NEED FERTILIZATION TO COMPLETE MEIOSIS; SPERM COMPLETE MEIOSIS WITHOUT FERTILIZATION

All amphimictic metazoans require sperm for the completion of female meiosis; however, cnidarians and sea urchins can accomplish this process without the need for plasma (Gorelick, 2012). Moreover, nearly all other phyla, such as arthropods, nematodes, rotifers, *Platyhelminthes*, mollusks, and annelids, as well as chordates (such as fish, reptiles, and birds), exhibit automictic parthenogenesis (e.g., centralized or terminal fusion).

It is evident that sperm fertilization is not necessary for these automictic eggs to complete meiosis. In order for the female haploid phase (megagametophyte) of plants (*Archaeplastida*), along with other heterosporous organisms, like some *Stramenopiles* (e.g., Laminaria), to have multicellular stages, female meiosis must occur independently of sperm.

5.13.1.2 EGG NUCLEI HAVE NUCLEAR PORES; SPERM NUCLEI LACK NUCLEAR PORES

Nucleotide mRNA and proteins can move freely in and out of a nucleus thanks to complex structures called nuclear pores, which are based on certain transmembrane proteins. Considering that all somatic cells have pores in their nuclear membranes, haploid male nuclei are the only types of nuclei that are totally devoid of them. It appears from our little examination of metazoans that not all sperm are devoid of nuclear pores. To be more precise, only the nuclei of sperm are devoid of nuclear pores or only present in certain locations. Numerous phyla of animals, including certain chordates and insects, include instances of localized nuclear holes in their sperm. In numerous taxa, nuclear pores undergo redistribution and localization in the "redundant nuclear envelope" located at the posterior (flagellar) ends in the sperm during spermatogenesis. The surplus nuclear material that is stored and packaged close to the sperm tail during nuclear condensation—which does not encompass chromosomes—is most likely what the redundant nuclear envelope is used for.

In contrast to animals, flowering plants—at least the few taxa that have been studied—have nuclear holes in their sperm cells. Nuclear pores were found in the cells that produce sperm of the eudicots Brassica and Plumbago, respectively, as reported by Taylor et al. (1991).

These nuclear pores do not seem to be confined. According to a number of studies, nuclear pores are compatible with the transcription and translation that monocot sperm cells in *Oryza, Zea,* and *Lilium* experience (Abiko et al., 2013; Blomstedt, Knox, and Singh, 1996; Zhang, Gifford, and Cass, 1993; Zhao, Yang, and Wang, 2013).

During the formation of sperm (spermatozoa), the monophyte *Marsilea* employs reserved mRNA transcripts for the synthesis of proteins, which is circumstantial evidence of the absence of nuclear pores. *Marsilea* gametes, therefore, have metazoan-like nuclear pores.

Given that the sperm of flowering plants contain nuclear pores, there is no reason for them to replace this nuclear membrane following plasmogamy. On the other hand, *Marsilea*, which we expect to have sperm that lack nuclear pores, probably does undergo metazoan-like replacement of their nuclear envelope following plasmogamy so that the male nucleus can replicate its chromosomes prior to karyogamy (Gorelick and Carpinone, 2009).

5.13.1.3 HAPLOID FEMALES ARE PRODUCED BY ASYMMETRICAL MEIOSIS; HAPLOID MALES ARE PRODUCED BY SYMMETRICAL MEIOSIS

Any meiotic division that results in the production of female (male) nuclei that are haploid, as in female (male) meiosis. In other words, the gametophyte(s) that result produce eggs during female meiosis, while the gametophytes that result produce sperm during male meiosis. All metazoans undergo asymmetric female meiosis, according to our earlier hypothesis, in which only one (or two, with middle or terminal fusion) of the developing haploid nuclei is passed on to the following generations (Gorelick et al., 2013).

It is believed that in order to improve the fitness of the resultant embryo, eggs require greater nutritional support than sperm (Schuh and Ellenberg, 2008). Additionally, we had previously proposed that all metazoans undergo symmetric male meiosis, with each of the four gametic nuclei that emerge having the capacity to pass on information to the following generation.

Meiosis that results in females appears to be asymmetrical in all mammals. A few meiotic products in females are smaller than others; these smaller cells and nuclei are eliminated as polar bodies, including polar nuclei, via evolution. Only one of the four haploid products produced by female meiosis seems to be passed on to the following generation in amphimictic metazoans, and it can only do so if it is nourished by another gamete.

Similar to this, in those taxa where pre-meiotic duplication and gamete doubling occur, only the largest haploid female nucleus endures until the following generation. As polar bodies or polar nuclei, the other two haploid byproducts of meiosis in core or terminal fusing are expelled or broken down. Beyond the observation that certain organisms with amoeboid sperm appear to produce five cells by meiosis, but only four of them have nuclei, we have not uncovered any further exceptions to this norm in metazoans (Smith, 2006).

Meiosis creates a tetrad of four identically sized and functional megaspores in the two genera of heterosporous lycophytes that are currently in existence: *Selaginella* and *Isoetes* (Blackmore et al., 2012; Eames, 1936; Morbelli et al., 2003). According to Blackmore et al. (2012), these four female meiotic products are somewhat unusual in that they are substantially larger than the diploid cell from which they originated. The products of male meiosis in lycophytes and zoidogamous seed plants also share this characteristic (Blackmore et al., 2012; Friedman and Gillford, 1997; Norstog, 1993; Poort et al., 1996; Stewart, 1951).

The mature megagametophyte and mature megaspore in current heterosporous lycophytes are of the same size. Importantly, in heterosporous lycophytes, there is no disagreement over the identity of female or male gametophytes. Many tiny sperm are produced by male lycophyte gametophytes, which swim to egg cells with flagella. Large sessile egg cells are attached to the female gametophytes of lycophytes, and following the fertilization process, the embryo invades the female gametophyte. Despite symmetrical meiosis in both sexes, female lycophyte gametophytes are greater than male gametophytes (classical anisogamy).

One monocot and one eudicot flowering plant lineage, at least, have developed asymmetrical male meiosis secondary to other evolutionary processes. Some species in the heather family *Ericaceae* subfamily *Styphelioideae* and the sedge family *Cyperaceae* produce so-called "pseudomonad" pollen, in which three of the four products of male meiosis degenerate much like polar bodies in nature (Brown and Lemmon, 2000; Furness, 2009; Furness and Rudall, 2011; Kirpes et al., 1996; San Martin et al., 2013).

Of all four outcomes of male meiosis, the pseudomonad granule of pollen is the largest; the cytoplasm of the other three haploid cells is smaller. It is assumed that pseudomonad pollen production results from uneven chromosomal segregation during meiosis, either because these species have holocentric chromosomes or dispersed centromeres (Furness and Rudall, 2011; Hipp, 2007).

Of all four outcomes of male meiosis, the pseudomonad granule of pollen is the largest; the cytoplasm of the other three haploid cells is smaller. It is assumed that pseudomonad pollen production results from uneven chromosomal segregation during meiosis, either because these species have holocentric chromosomes or dispersed centromeres (Furness and Rudall, 2011; Hipp, 2007). However, in each meiotic division, the single pseudomonad pollen grain usually evolves into a completely typical angiosperm pollen tube containing two sperm cells. The female meiosis of the few angiosperms that have asymmetrical male meiosis is asymmetrical, much like that of conventional angiosperms.

5.14 ANGIOSPERM

For the survival of a species, an essential aspect of life for all living things is reproduction. The process in multicellular eukaryotic creatures starts and primarily involves the specification of the meiocyte, a cell specialized in producing germ-line cells. By designating different types of cells with a meiotic fate, eukaryotic organisms are able to avoid genome doubling after every reproduction event and maintain genetic variety. Prior to the creation of gametes, the meiocyte's chromosome set is divided in half during meiosis, and genetic material from the maternal as well as paternal genomes recombines.

Plants go through two alternate generations in their life cycle: the haploid gametophytic phase and the diploid sporophytic phase (Friedman, 2013; Pandey et al., 2021).

Meiosis happens shortly after the sporophytic generation forms during syngamy, or fertilization, meaning that the majority of the organism's lifespan in these plant lineages is devoted to the gametophyte generation. In lower plants, like mosses, the gametophyte is an independent form that develops independently of the sporophyte. On the other hand, flowering plants live most of their lives as sporophytes. Angiosperm species' mature gametophytes are identified as being embedded in sporophyte tissue.

Both the female and male floral organs, an archesporial cell, a progenitor of meiocyte, differentiate from a specific group of somatic cells upon the start of the shift to the gametophytic phase. In the male floral organs, the anther primordium's four corners' hypodermal cells differentiate into archesporial cells, which then go through periclinal division to produce an outer primary parietal cell lineage and an inner primary sporogenous cell lineage (Ren et al., 2018; Sanders et al., 1999). The archesporial cell in female floral organs develops from a single hypodermal cell at the top of the ovule primordia and, without further cell divisions, instantly transforms into the female meiocyte known as the megaspore mother cell (MMC).

In the female meiosis of angiosperms, asymmetry can be overt or inconspicuous (Hoekstra, 2011). Nuclei may have an instantaneous or delayed destiny. Angiosperms that undergo tetraploid formation of female gametophytes lack polar bodies or polar nuclei, but they nevertheless reject three of the four meiotic products. The so-called "germinal spore," which has a single haploid nucleus (Haig, 1986), develops into the fertilized egg cell that gives rise to the zygote after fertilization. The two distinct polar nuclei of the basal cell, which become triploid endosperm following multiple fertilization, are eventually formed by the mitotic division of a second haploid nucleus. It should be noted that there are issues with standard nomenclature in this case because the term "polar nucleus" is used in two separate contexts. In fact, angiosperm female gametophytes vary greatly throughout species (Klekowski, 1988; Williams and Friedman, 2002; Yadegari and Drews, 2004).

Endosperm containing genomes of different angiosperms is not inherited by subsequent generations. The female gametophyte comprises the other two haploid products that result from meiosis, but neither the egg nor the central cell does.

5.15 GYMNOSPERM

Known as "simple polyembryony," it occurs in many different groups of gymnosperm species that contain one or more eggs per megagametophyte fertilized by different related sperms, producing numerous diploid embryos in only one ovule (Becwar et al., 1991; Singh, 1978; Smoot and Taylor, 1986). With the exception of Gnetales, all gymnosperms appear to be monospermic, meaning that these numerous fertilized eggs solely derive from the chalazal product of female meiosis (Friedman, 1990, 2015; Carmichael and Friedman, 1995). Therefore, asymmetrical female meiosis is still present in these aberrant gymnosperm situations (Nagasato et al., 2001). Similar to land algae-containing plants, one of the brown algae, *Fucales*, can have about one to eight haploid nuclei in each egg cell. This is because meiotic as well as mitotic nuclear division, or karyogenesis, does not always coincide with CD, or cytogenesis. *Fucales* generate a maximum of eight fertilized eggs, depending on how many of the four haploid nuclei that arise from female meiosis disintegrate as polar nuclei after a single mitotic split of each haploid nucleus (Brawley et al., 1977; Nagasato et al., 2001).

5.15.1 KELPS

Even though most kelp are heterosporous, they only undergo one type of meiosis—a meiotic division that yields about 50% male spores and 50% female spores. Put another way, referring to male or female meiosis in the majority of cases is illogical.

It is challenging to deduce lifecycles from fossil gametes since they are frequently separated from both their sporophytes and gametophytes. It is challenging to determine if a single division of meiosis produces both female and male spores in fossil taxa, much less to infer anisogamy from fossils (Cichan et al., 1984). According to Bell and Hemsley (2000), the majority of monilophytes, certain lycophytes, and all living anthocerotophytes are homosporous. First, we defined female (male) meiosis as any meiotic division that results in the production of female (male) haploid nuclei. As far as we know, meiosis that results in two female and two male spores that are haploid never forms polar bodies.

5.16 MEIOSIS IN DIFFERENT PLANTS

Meiosis in plants can speed up agricultural applications by providing a foundation for advancements in genetics, breeding, reproduction, and fertility (Sanchez-Moran et al., 2008).

Because meiosis in plants continues until the end of the tetrad stage of development, with meiotic defects such as massive chromosome segregation, plants are considered a suitable model system for studying meiosis. Unlike mammals, there are no disconcerting consequences from the onset of meiosis arrest and apoptosis (Higgins et al., 2004; Mercier and Grelon, 2008).

In a larger sense, the kingdom Plantae, also known as *Archaeplastida*, is made up of land plants. The most exclusive group, known as chlorophytes, as well as charophytes, are freshwater unicellular algae, and photoautotrophic red algae are known as rhodophytes. According to Puttick et al. (2018), terrestrial plants can be further divided into spermatophytes, which include gymnosperms and angiosperms, bryophytes, lycophytes, and pteridophytes, which include ferns.

Compared to the total of 5400 mammalian species, land plants are thought to comprise about 500,000 species, indicating that they are highly varied (Corlett, 2016). The majority of research on meiosis in plants has been done on angiosperms, and the model organism *A. thaliana*, as well as rice, maize, wheat, and barley, are the subjects of most studies characterizing meiotic genes. Approximately 100 genes related to meiosis in *A. thaliana* have been investigated functionally to date (Zhang et al., 2018). But given the broad range of plants, understanding the development of meiosis in this kingdom will require more than a cursory examination of a few angiosperm models.

KEYWORDS

- ***Arabidopsis thaliana***
- **genetic recombination**
- **heterosporous plants**
- **homosporous plants**
- **meiosis**
- **plant reproduction**
- **tetrad stage**

REFERENCES

Aggarwal, D. D., Rashkovetsky, E., Michalak, P., Cohen, I., Ronin, Y., Zhou, D., & Korol, A. B. (2015). Experimental evolution of recombination and crossover interference in *Drosophila* caused by directional selection for stress-related traits. *BMC Biology, 13*, 1–14.

Ahmed, S., Cock, J. M., Pessia, E., Luthringer, R., Cormier, A., Robuchon, M., & Coelho, S. M. (2014). A haploid system of sex determination in the brown alga *Ectocarpus* sp. *Current Biology, 24*(17), 1945–1957.

Allen, C. E. (1945). The genetics of bryophytes. II. *The Botanical Review, 11*, 260–287.
Anderson, L. K., Lohmiller, L. D., Tang, X., Hammond, D. B., Javernick, L., Shearer, L., & Falque, M. (2014). Combined fluorescent and electron microscopic imaging unveils the specific properties of two classes of meiotic crossovers. *Proceedings of the National Academy of Sciences, 111*(37), 13415–13420.
Archetti, M. (2003). A selfish origin for recombination. *Journal of Theoretical Biology, 223*(3), 335–346.
Archetti, M. (2010). Complementation, genetic conflict, and the evolution of sex and recombination. *Journal of Heredity, 101*(Suppl_1), S21–S33.
Arkhipova, I., & Meselson, M. (2005). Deleterious transposable elements and the extinction of asexuals. *BioEssays, 27*(1), 76–85.
Ash-King, M. (2013). *Challenging Popular Myths of Sex, Gender and Biology.* Springer Science & Business Media.
Bachtrog, D., Kirkpatrick, M., Mank, J. E., McDaniel, S. F., Pires, J. C., Rice, W., & Valenzuela, N. (2011). Are all sex chromosomes created equal? *Trends in Genetics, 27*(9), 350–357.
Bateman, R. M., & DiMichele, W. A. (1994). Heterospory: The most iterative key innovation in the evolutionary history of the plant kingdom. *Biological Reviews, 69*(3), 345–417.
Baudat, F., Imai, Y., & De Massy, B. (2013). Meiotic recombination in mammals: Localization and regulation. *Nature Reviews Genetics, 14*(11), 794–806.
Becwar, M. R., Blush, T. D., Brown, D. W., & Chesick, E. E. (1991). Multiple paternal genotypes in embryogenic tissue derived from individual immature loblolly pine seeds. *Plant Cell, Tissue and Organ Culture, 26*, 37–44.
Bell, G. (1978). The evolution of anisogamy. *Journal of Theoretical Biology, 73*(2), 247–270.
Bell, P. R., & Hemsley, A. R. (2000). *Green Plants: Their Origin and Diversity.* Cambridge University Press.
Berchowitz, L. E., Francis, K. E., Bey, A. L., & Copenhaver, G. P. (2007). The role of AtMUS81 in interference-insensitive crossovers in *A. thaliana. Genetics, 3*(8), e132.
Bernstein, H., & Bernstein, C. (2010). Evolutionary origin of recombination during meiosis. *BioScience, 60*(7), 498–505.
Bischler, H. (1986). *Marchantia polymorpha* LS LAT. Karyotype analysis. *The Journal of the Hattori Botanical Laboratory, 60*, 105–117.
Bishop, D. K., Park, D., Xu, L., & Kleckner, N. (1992). DMC1: A meiosis-specific yeast homolog of *E. coli* recA required for recombination, synaptonemal complex formation, and cell cycle progression. *Cell, 69*(3), 439–456.
Bomfleur, B., McLoughlin, S., & Vajda, V. (2014). Fossilized nuclei and chromosomes reveal 180 million years of genomic stasis in royal ferns. *Science, 343*(6177), 1376–1377.
Carmichael, J. S., & Friedman, W. E. (1995). Double fertilization in *Gnetum gnemon*: The relationship between the cell cycle and sexual reproduction. *The Plant Cell, 7*(12), 1975–1988.
Cavalier-Smith, T. (2002). Origins of the machinery of recombination and sex. *Heredity, 88*(2), 125–141.

Chelysheva, L., Vezon, D., Chambon, A., Gendrot, G., Pereira, L., Lemhemdi, A., & Grelon, M. (2012). The *Arabidopsis* HEI10 is a new ZMM protein related to Zip3. *PLoS Genetics, 8*(7), e1002799.

Chmátal, L., Gabriel, S. I., Mitsainas, G. P., Martínez-Vargas, J., Ventura, J., Searle, J. B., & Lampson, M. A. (2014). Centromere strength provides the cell biological basis for meiotic drive and karyotype evolution in mice. *Current Biology, 24*(19), 2295–2300.

Choi, K., & Henderson, I. R. (2015). Meiotic recombination hotspots: A comparative view. *The Plant Journal, 83*(1), 52–61.

Choulet, F., Alberti, A., Theil, S., Glover, N., Barbe, V., Daron, J., & Feuillet, C. (2014). Structural and functional partitioning of bread wheat chromosome 3B. *Science, 345*(6194), 1249721.

Cichan, M. A., & Taylor, T. N. (1984). Ultrastructural studies of *in situ* Devonian spores: *Protobarinophyton pennsylvanicum* Brauer. *Review of Palaeobotany and Palynology, 41*(3, 4), 167–175.

Corlett, R. T. (2016). Plant diversity in a changing world: Status, trends, and conservation needs. *Plant Diversity, 38*(1), 10–16.

Demirci, S., van Dijk, A. D., Sanchez Perez, G., Aflitos, S. A., de Ridder, D., & Peters, S. A. (2015). Distribution, position, and genomic characteristics of crossovers in tomato recombinant inbred lines derived from an interspecific cross between *Solanum lycopersicum* and *Solanum pimpinellifolium*. *The Plant Journal, 89*(3), 554–564.

Ferdous, M., Higgins, J. D., Osman, K., Lambing, C., Roitinger, E., Mechtler, K., & Franklin, F. C. H. (2012). Inter-homolog crossing-over and synapsis in *Arabidopsis* meiosis are dependent on the chromosome axis protein AtASY3. *PLoS Genetics, 8(2)*, e1002507.

Friedman, W. E. (1990). Sexual reproduction in *Ephedra nevadensis* (Ephedraceae): Further evidence of double fertilization in a nonflowering seed plant. *American Journal of Botany, 77*(12), 1582–1598.

Friedman, W. E. (2013). One genome, two ontogenies. *Science, 339*(6123), 1045, 1046.

Furness, C. A., & Rudall, P. J. (2011). Selective microspore abortion correlated with aneuploidy: An indication of meiotic drive. *Sexual Plant Reproduction, 24*, 1–8.

Fyon, F., Cailleau, A., & Lenormand, T. (2015). Enhancer runaway and the evolution of diploid gene expression. *PLoS Genetics, 11*(11), e1005665.

Girard, C., Crismani, W., Froger, N., Mazel, J., Lemhemdi, A., Horlow, C., & Mercier, R. (2014). FANCM-associated proteins MHF1 and MHF2, but not the other Fanconi anemia factors, limit meiotic crossovers. *Nucleic Acids Research, 42*(14), 9087–9095.

Gorelick, R. (2003). Evolution of dioecy and sex chromosomes via methylation driving Muller's ratchet. *Biological Journal of the Linnean Society, 80*(2), 353–368.

Gorelick, R. (2012). Meiosis is not gender neutral. *BioScience, 62*(7), 623–624.

Gorelick, R. (2015). Theory for why dioecious plants have equal length sex chromosomes. *American Journal of Botany, 92*(6), 979–984.

Griffiths, S., Sharp, R., Foote, T. N., Bertin, I., Wanous, M., Reader, S., & Moore, G. (2006). Molecular characterization of Ph1 as a major chromosome pairing locus in polyploid wheat. *Nature, 439*(7077), 749–752.

Haig, D. (1986). Conflicts among megaspores. *Journal of Theoretical Biology, 123*(4), 471–480.

Haig, D. (1993). Alternatives to meiosis: The unusual genetics of red algae, microsporidia, and others. *Journal of Theoretical Biology, 163*(1), 15–31.
Haig, D. (2010). Games in tetrads: Segregation, recombination, and meiotic drive. *The American Naturalist, 176*(4), 404–413.
Haig, D. (2014). Coadaptation and conflict, misconception and muddle, in the evolution of genomic imprinting. *Heredity, 113*(2), 96–103.
Haig, D., & Grafen, A. (1991). Genetic scrambling as a defense against meiotic drive. *Journal of Theoretical Biology, 153*(4), 531–558.
Heard, E., & Martienssen, R. A. (2014). Transgenerational epigenetic inheritance: Myths and mechanisms. *Cell, 157*(1), 95–109.
Herskowitz, I. (1988). Life cycle of the budding yeast *Saccharomyces cerevisiae*. *Microbiological Reviews, 52*(4), 536–553.
Higgins, J. D., Armstrong, S. J., Franklin, F. C. H., & Jones, G. H. (2004). The *Arabidopsis* MutS homolog AtMSH4 functions at an early step in recombination: Evidence for two classes of recombination in *Arabidopsis*. *Genes & Development, 18*(20), 2557–2570.
Higgins, J. D., Buckling, E. F., Franklin, F. C. H., & Jones, G. H. (2008). Expression and functional analysis of AtMUS81 in *Arabidopsis* meiosis reveals a role in the second pathway of crossing-over. *The Plant Journal, 54*(1), 152–162.
Hipp, A. L. (2007). Nonuniform processes of chromosome evolution in sedges (*Carex*: Cyperaceae). *Evolution, 61*(9), 2175–2194.
Hiruta, C., Nishida, C., & Tochinai, S. (2010). Abortive meiosis in the oogenesis of parthenogenetic *Daphnia pulex*. *Chromosome Research, 18*, 833–840.
Holliday, R. (1984). The biological significance of meiosis. In *Symposia of the Society for Experimental Biology* (Vol. *38*, pp. 381–394).
Hong, S., Sung, Y., Yu, M., Lee, M., Kleckner, N., & Kim, K. P. (2013). The logic and mechanism of homologous recombination partner choice. *Molecular Cell, 51*(4), 440–453.
Hunter, N. (2015). Meiotic recombination: The essence of heredity. *Cold Spring Harbor Perspectives in Biology, 7*(12), a016618.
Hurst, L. D. (1993). Drunken walk of the diploid. *Nature, 365*(6443), 206–207.
Ingouff, M., Selles, B., Michaud, C., Vu, T. M., Berger, F., Schorn, A. J., & Grimanelli, D. (2017). Live-cell analysis of DNA methylation during sexual reproduction in *Arabidopsis* reveals context and sex-specific dynamics controlled by noncanonical RdDM. *Genes & Development, 31*(1), 72–83.
Jacob, P., Avni, A., & Bendahmane, A. (2018). Translational research: Exploring and creating genetic diversity. *Trends in Plant Science, 23*(1), 42–52.
Khanna, K. R. (1971). Sex chromosome bryophytes. *Nucleus, 14*(1), 14.
Kleckner, N. (1996). Meiosis: How could it work? *Proceedings of the National Academy of Sciences, 93*(16), 8167–8174.
Kleckner, N. (2006). Chiasma formation: Chromatin/axis interplay and the role(s) of the synaptonemal complex. *Chromosoma, 115*(3), 175–194.
Klekowski, E. J. (1988). *Mutation, Developmental Selection, and Plant Evolution.*
Lambing, C., Franklin, F. C. H., & Wang, C. J. R. (2017). Understanding and manipulating meiotic recombination in plants. *Plant Physiology, 173*(3), 1530–1542.
Lambing, C., Osman, K., Nuntasoontorn, K., West, A., Higgins, J. D., Copenhaver, G. P., & Franklin, F. C. H. (2015). *Arabidopsis* PCH2 mediates meiotic chromosome remodeling and maturation of crossovers. *PLoS Genetics, 11*(7), e1005372.

Lee, D. H., Kao, Y. H., Ku, J. C., Lin, C. Y., Meeley, R., Jan, Y. S., & Wang, C. J. R. (2015). The axial element protein DESYNAPTIC2 mediates meiotic double-strand break formation and synaptonemal complex assembly in maize. *The Plant Cell, 27*(9), 2516–2529.

Lemmon, B. E. (2000). The cytoskeleton and polarization during pollen development in *Carex blanda* (Cyperaceae). *American Journal of Botany, 87*(1), 1–11.

Lenormand, T., Engelstädter, J., Johnston, S. E., Wijnker, E., & Haag, C. R. (2016). Evolutionary mysteries in meiosis. *Philosophical Transactions of the Royal Society B: Biological Sciences, 371*(1706), 20160001.

Li, X., Li, L., & Yan, J. (2015). Dissecting meiotic recombination based on tetrad analysis by single-microspore sequencing in maize. *Nature Communications, 6*(1), 6648.

Luo, Q., Li, Y., Shen, Y., & Cheng, Z. (2014). Ten years of gene discovery for meiotic event control in rice. *Journal of Genetics and Genomics, 41*(3), 125–137.

Ma, W., Gabriel, T. S., Martis, M. M., Gursinsky, T., Schubert, V., Vrána, J., & Houben, A. (2017). Rye B chromosomes encode a functional Argonaute-like protein with *in vitro* slicer activities similar to its A chromosome paralog. *New Phytologist, 213*(2), 916–928.

Mable, B. K., & Otto, S. P. (1998). The evolution of life cycles with haploid and diploid phases. *BioEssays, 20*(6), 453–462.

Madlung, A., Tyagi, A. P., Watson, B., Jiang, H., Kagochi, T., Doerge, R. W., & Comai, L. (2015). Genomic changes in synthetic *Arabidopsis* polyploids. *The Plant Journal, 41*(2), 221–230.

Mercier, R., & Grelon, M. (2008). Meiosis in plants: Ten years of gene discovery. *Cytogenetic and Genome Research, 120*(3, 4), 281–290.

Mercier, R., Mézard, C., Jenczewski, E., Macaisne, N., & Grelon, M. (2015). The molecular biology of meiosis in plants. *Annual Review of Plant Biology, 66*, 297–327.

Michod, R. E., & Levin, B. R. (1988). *The Evolution of Sex: An Examination of Current Ideas.*

Mira, A. (1998). Why is meiosis arrested? *Journal of Theoretical Biology, 194*(2), 275–287.

Nagasato, C., Motomura, T., & Ichimura, T. (2001). Degeneration and extrusion of nuclei during oogenesis in *Silvetia babingtonii, Cystoseira hakodatensis*, and *Sargassum confusum* (Fucales, Phaeophyceae). *Phycologia, 40*(5), 411–420.

Neale, M. J., Pan, J., & Keeney, S. (2005). Endonucleolytic processing of covalent protein-linked DNA double-strand breaks. *Nature, 436*(7053), 1053–1057.

Obeso, D., Pezza, R. J., & Dawson, D. (2014). Couples, pairs, and clusters: Mechanisms and implications of centromere associations in meiosis. *Chromosoma, 123*, 43–55.

Osman, K., Higgins, J. D., Sanchez-Moran, E., Armstrong, S. J., & Franklin, F. C. H. (2011). Pathways to meiotic recombination in *Arabidopsis thaliana. New Phytologist, 190*(3), 523–544.

Osman, K., Yang, J., Roitinger, E., Lambing, C., Heckmann, S., Howell, E., & Franklin, F. C. H. (2018). Affinity proteomics reveals extensive phosphorylation of the *Brassica* chromosome axis protein ASY1 and a network of associated proteins at prophase I of meiosis. *The Plant Journal, 93*(1), 17–33.

Otto, S. P. (2009). The evolutionary enigma of sex. *The American Naturalist, 174*(S1), S1–S14.

Otto, S. P., & Gerstein, A. C. (2006). Why have sex? The population genetics of sex and recombination. *Biochemical Society Transactions, 34*(4), 519–522.

Otto, S. P., & Goldstein, D. B. (1992). Recombination and the evolution of diploidy. *Genetics, 131*(3), 745–751.

Ottolini, C. S., Newnham, L. J., Capalbo, A., Natesan, S. A., Joshi, H. A., Cimadomo, D., & Hoffmann, E. R. (2015). Genome-wide maps of recombination and chromosome segregation in human oocytes and embryos show selection for maternal recombination rates. *Nature Genetics, 47*(7), 727–735.

Pandey, S., Moradi, A. B., Dovzhenko, O., Touraev, A., Palme, K., & Welsch, R. (2022). Molecular control of sporophyte-gametophyte ontogeny and transition in plants. *Frontiers in Plant Science, 12*, 789789.

Panizza, S., Mendoza, M. A., Berlinger, M., Huang, L., Nicolas, A., Shirahige, K., & Klein, F. (2011). Spo11-accessory proteins link double-strand break sites to the chromosome axis in early meiotic recombination. *Cell, 146*(3), 372–383.

Parfrey, L. W., Lahr, D. J., & Katz, L. A. (2008). The dynamic nature of eukaryotic genomes. *Molecular Biology and Evolution, 25*(4), 787–794.

Phillips, D., Wnetrzak, J., Nibau, C., Barakate, A., Ramsay, L., Wright, F., & Jenkins, G. (2013). Quantitative high-resolution mapping of HvMLH3 foci in barley pachytene nuclei reveals a strong distal bias and weak interference. *Journal of Experimental Botany, 64*(8), 2139–2154.

Puttick, M. N., Morris, J. L., Williams, T. A., Cox, C. J., Edwards, D., Kenrick, P., & Donoghue, P. C. (2018). The interrelationships of land plants and the nature of the ancestral embryophyte. *Current Biology, 28*(5), 733–745.

Ramesh, M. A., Malik, S. B., & Logsdon, J. M. (2005). A phylogenomic inventory of meiotic genes: Evidence for sex in *Giardia* and an early eukaryotic origin of meiosis. *Current Biology, 15*(2), 185–191.

Rescan, M., Lenormand, T., & Roze, D. (2016). Interactions between genetic and ecological effects on the evolution of life cycles. *The American Naturalist, 187*(1), 19–34.

Robert, T., Nore, A., Brun, C., Maffre, C., Crimi, B., Guichard, V., & De Massy, B. (2016). The TopoVIB-Like protein family is required for meiotic DNA double-strand break formation. *Science, 351*(6276), 943–949.

Rusche, L. N., & Rine, J. (2010). Switching the mechanism of mating type switching: A domesticated transposase supplants a domesticated homing endonuclease. *Genes & Development, 24*(1), 10–14.

San Martin, J. A. B., de Jesus Andrade, C. G. T., Mastroberti, A. A., de Araújo Mariath, J. E., & Vanzela, A. L. (2013). Asymmetric cytokinesis guides the development of pseudomonads in *Rhynchospora pubera* (Cyperaceae). *Cell Biology International, 37*(3), 203–212.

Sanchez-Moran, E., Osman, K., Higgins, J. D., Pradillo, M., Cunado, N., Jones, G. H., & Franklin, F. C. H. (2008). ASY1 coordinates early events in the plant meiotic recombination pathway. *Cytogenetic and Genome Research, 120*(3, 4), 302–312.

Schubert, V. (2017). Super-resolution microscopy–applications in plant cell research. *Frontiers in Plant Science, 8*, 531.

Serra, H., Lambing, C., Griffin, C. H., Topp, S. D., Nageswaran, D. C., Underwood, C. J., & Henderson, I. R. (2018). Massive crossover elevation via combination of HEI10 and *recq4a recq4b* during *Arabidopsis* meiosis. *Proceedings of the National Academy of Sciences, 115*(10), 2437–2442.

Singh, H. (1978). *Embryology of Gymnosperms.* https://d-nb.info/780421256/04 (accessed on 18 March 2025).

Smith, J. M., & Maynard-Smith, J. (1978). *The Evolution of Sex* (Vol. 4). Cambridge: Cambridge University Press.

Smoot, E. L., & Taylor, T. N. (1986). Evidence of simple polyembryony in Permian seeds from Antarctica. *American Journal of Botany, 73*(7), 1079–1081.

Souza-Júnior, S. A., Becker, T. C. A., & Castro-Prado, M. A. A. (2007). Asexual recombination in a *uvsH* mutant of *Aspergillus nidulans*. *Biological Research, 40*(1), 65–71.

Speijer, D., Lukeš, J., & Eliáš, M. (2015). Sex is a ubiquitous, ancient, and inherent attribute of eukaryotic life. *Proceedings of the National Academy of Sciences, 112*(29), 8827–8834.

Togashi, T., & Cox, P. A. (2011). The evolution of anisogamy: A fundamental phenomenon underlying sexual selection. *The Quarterly Review of Biology, 87*(2).

Ubeda, F., & Haig, D. (2005). On the evolutionary stability of Mendelian segregation. *Genetics, 170*(3), 1345–1357.

Villeneuve, A. M., & Hillers, K. J. (2001). Whence meiosis? *Cell, 106*(6), 647–650.

Wang, K., Wang, M., Tang, D., Shen, Y., Qin, B., Li, M., & Cheng, Z. (2011). PAIR3, an axis-associated protein, is essential for the recruitment of recombination elements onto meiotic chromosomes in rice. *Molecular Biology of the Cell, 22*(1), 12–19.

Wang, M., Wang, K., Tang, D., Wei, C., Li, M., Shen, Y., & Cheng, Z. (2010). The central element protein ZEP1 of the synaptonemal complex regulates the number of crossovers during meiosis in rice. *The Plant Cell, 22*(2), 417–430.

Wang, Y., & Copenhaver, G. P. (2018). Meiotic recombination: Mixing it up in plants. *Annual Review of Plant Biology, 69*, 577–609.

Wang, Y., Jha, A. K., Chen, R., Doonan, J. H., & Yang, M. (2010). Polyploidy-associated genomic instability in *Arabidopsis thaliana*. *Genesis, 48*(4), 254–263.

Wijnker, E., & Schnittger, A. (2013). Control of the meiotic cell division program in plants. *Plant Reproduction, 26*, 143–158.

Wilkins, A. S., & Holliday, R. (2009). The evolution of meiosis from mitosis. *Genetics, 181*(1), 3–12.

Williams, J. H., & Friedman, W. E. (2002). Identification of diploid endosperm in an early angiosperm lineage. *Nature, 415*(6871), 522–526.

Yadegari, R., & Drews, G. N. (2004). Female gametophyte development. *The Plant Cell, 16*(suppl_1), S133–S141.

Yelina, N. E., Choi, K., Chelysheva, L., Macaulay, M., De Snoo, B., Wijnker, E., & Henderson, I. R. (2012). Epigenetic remodeling of meiotic crossover frequency in *Arabidopsis thaliana*. *PLoS Genet, 8*(8), e1002844.

Zhang, L., Kong, H., Ma, H., & Yang, J. (2018). Phylogenomic detection and functional prediction of genes potentially important for plant meiosis. *Gene, 643*, 83–97.

Zhang, L., Liang, Z., Hutchinson, J., & Kleckner, N. (2014). Crossover patterning by the beam-film model: Analysis and implications. *Genetics, 10*(1), e1004042.

CHAPTER 6

Structural Changes in Chromosomes

MOMINA SAEED,[1] ABIDA PARVEEN,[1,*] SHAGUFTA PERVEEN,[1] SARA ZAFAR,[1] QAISAR MAHMOOD,[2,3] and SUMAIRA THIND[1]

[1]*Department of Botany, Government College University, Faisalabad, Pakistan*

[2]*Department of Biology, College of Sciences, Sakhir, Bahrain*

[3]*Department of Environmental Sciences, Comsats University, Islamabad, Abbottabad Campus, Pakistan*

*Corresponding author

ABSTRACT

A genome's dimension, the center and nucleolar organizer locations, and chromomeres, a form of hetero and ring arrangements, can all be used to architecturally identify its structure. With a magnifying glass, cytological examination can identify four different forms of chromosomal structural deviations: modifications, redundant data, inverted shapes, and mutations. Nonetheless, certain modifications are too minute to be identified using cytology. Eliminations are parts of genomes that are missing. While intermediate losses may carry inheritable genes that are revealed by loss and can be either nonlethal or deadly, unilateral lesions have the potential to be fatal. Repetition can result in asymmetry in genetic material, which can affect the appearance of an organism and increase the diversity of gene activities. A chromosomal segment's 180° turn results in inverted shapes. This, in pure conditions, presents minimal difficulty to an organism; nevertheless, inverted heterozygotes frequently experience matching issues throughout meiosis, resulting in the creation of reversal loops. Appealing materials are typically produced when loops cross over. In pericentric and

Chromosomal and Genomic Structures and Functions,Volume 1: Chromosomes and Cell Division.
Tariq Ahmad Bhat and Jameel M. Al-Khayri (Eds)

DOI: 10.1201/9781779640253-6

paracentric inverted postures, there will be differences in the overlap products. Moving a cytogenetic fragment to a different site in the genetic code is known as a relocation. Cognitive impairment and sterility are the two main health issues resulting from chromosomal reorganization in humans.

6.1 INTRODUCTION

It has been determined that essential activities taking place in the cell nucleus are significantly influenced by the three-dimensional (3D) arrangement of mammalian genomes. It has been demonstrated that the way genomes are folded in three dimensions is closely related to transcription, replication, and DNA damage and repair. Two complementary types of approaches provide the current understanding of 3D chromosomal structure. Initially and historically, there was microscopy, which unveiled several aspects of nuclear structure, including the presence of sub-nuclear organelles like the nucleolus (Pederson, 2011). Chromosome organization and location within the nucleus may be directly observed using fluorescence microscopy, which is its main benefit. However, even when recent advancements have significantly changed the range of options historically, its capabilities have been restricted in terms of throughput, genetic, and geographic resolution. We now have a better understanding of the chromatin fiber's fine-scale structure thanks to electron microscopy (Belmont Jr. et al., 2014). They depend on sequencing with high throughput and have transformed the study of nuclear anatomy by enabling high-resolution, genome-wide indicators of physical proximity events that occur within and between chromosomes. These intricate folding structures at the scale of Topologically associating domains (TADs) are largely influenced by compartmental contacts and polycomb-mediated linkages among activated and dormant genes (Mifsud et al., 2015). These measurements are typically understood in terms of chromosomal "contacts" or "interactions." Each chromosome is folded into intricate structural patterns that emerge at various sizes, according to 3C research. Genomic sequences that are active and inactive often interact exclusively to form A and B compartments (Lieberman-Aiden et al., 2009).

Topologically associating domains (TADs) are the domains whose boundaries are most preserved during cell development, although their precise definition is unclear because of the many mechanisms they arise from and the tremendous variety of sub-megabase structures of interaction (Zhan et al., 2017). Mammalian chromosomal complex folding appears to be influenced by polycomb-coated sequences and compartmental interactions. Experimental data indicating that chromosomal structure may be crucial in scaffolding the physical contacts between regulatory sequences, which are believed to help direct gene regulation during development and homeostasis, has generated great interest in the field as a result of these discoveries (Franke et al., 2016). Recent significant technical advancements have created previously unheard-of opportunities to investigate the molecular relationships between nuclear biology and chromosomal structure. The present paradigm will be expanded, improved measurement of dynamic molecular processes will be possible, and the main unanswered questions will be addressed by combining currently used methods (most notably 3C) with recently discovered approaches. In light of recent findings using novel and rapidly evolving orthogonal approaches, we address 3C methods and the conclusions they permitted in this review. We concentrate on the advantages, disadvantages, and potential benefits of combining different approaches, as well as the information they may provide on the biological significance and specifics of chromosomal structure to the true likelihoods of interaction between genomic regions. The interpretation of data is significantly affected by this (Imakaev et al., 2012).

Through indirect crosslinking processes caused by intervening nuclear proteins or even organelles, crosslinking may be able to capture connections that aren't caused by direct chemical reactions of the chromatin fiber. The joining may facilitate the creation of non-specific molecular hybrids between chromatin that is partly solubilized and cross-linked. Thus, it is crucial to compare and provide context for the 3C technique using orthogonal measures of genomic contacts that do not rely on bonding and fusion. The majority of 3C-based data at different genomic sizes are connected to colocalization patterns of FISH probes and/or inversely proportionate to their spatial distances, according to DNA FISH investigations, which employ crosslinking rather than ligation (Wang et al., 2016). Similar to the percentage of cells where two genes are closer than an established distance in DNA FISH, signals in 3C techniques originate from a particular population of cells whose two loci are physically close to one another. Generally speaking, this has an inverse relationship with the average spatial separation between the sensors. When multi-peak dual range dispersion is present, or when complex fluid procedures like ring protrusion are present (Fudenberg et al., 2017). Since the proportion of cells when the instruments are near to one another is greater than the median relative distance of the instruments, 3C readings should thus more closely coincide with FISH when compared with this fraction (Hastings et al., 2016). Furthermore, in DNA FISH, closeness is determined by subjective connect or separation boundaries that may not accurately reflect the real binding diameters in 3C, which unintentionally may vary throughout the entire genome based on factors like the quantity and affiliation of complexes of proteins attached to DNA (Bintu et al., 2018). Therefore, the findings of DNA FISH and Hi-C agree that two DNA sequences are spatially close when they are located within a couple of 100 nanometers of each other. This length span remains probably a number factor bigger than lengths, which is an intriguing observation. Necessary for mutual intimate interaction between complicated proteins attached to DNA. Nonetheless, the swift advancement of technology in all categories of approaches indicates that further understanding of these closer-range connections will be provided shortly (Hsieh et al., 2019). The presence of sections, TADs, and sub-TAD structures has been validated by a number of modern approaches oblique to both 3C and FISH, demonstrating that 3C methods correctly identify genomic connections happening within limited geographic ranges. It appears that such trends are not only the result of artifactual coupling of specific conversations, since "native" 4C and Hi-C conducted without coupling succeeded in identifying TAD borders and CTCF-mediated interaction (Cleynen et al., 2016).

Additional research using two ligation-free methods has demonstrated that contacts seen in 3C result from cytogenetic segments being close to one another (200 nm). When integrating the genome architecturally (GAM) (Beagrie et al., 2017), quantifying physical closeness (~200 nm) was achieved using extremely thin nucleus cryosectioning, and then large-scale sequencing of DNA isolated from central cuts, TADs, sections, and far-reaching inter-TAD connections were retrieved (Quinodoz et al., 2018). DNA fragments inside the identical crossing combination can be split-pool barcoded to distinguish between various association categories. Measurement correlation exists between short-term SPRITE interactions and far-reaching mediated bridging activities employing Hi-C data, which might potentially be controlled by subnuclear divisions. The last method to identify cytogenetic contacts in living cells via DNA adenine methylation is DamC, an altered form of DNA adenosine methyltransferase recognition (DamID) that is concurrently binding- and ligation-free (Redolfi et al., 2019).

The chromosome's actual form and structure inside the center of the cell are not fully specified by interaction frequencies alone. This presents a major obstacle for computational methods that use Hi-C data to infer 3D representations of chromatin or the entire nucleus, as well as for contrasting 3C-based information with microscopic evidence (in which distances that do not interact in reality are immediately noticed). To quantify the physical separations among DNA sequences and their locations in relation to nuclear features such as the nuclear shell and nucleoli, additional techniques are required. DamID can recognize Lamin-associated domains (LADs), which are DNA sections that frequently come into contact with the nuclear sheet (Peric-Hupkes et al., 2010). When antibody-coupled processors are directed toward noteworthy cytoplasmic characteristics (such as nuclear flecks), they produce diffuse biotinylated monomers that, in a distance-dependent fashion, mark neighboring DNA. The precise measurement of the lengths among subnuclear portions and genomic DNA is made possible by this method. Enhancements in imaging methods additionally make it possible to estimate distances within the nucleus more precisely. DNA FISH was previously restricted by poor biological recovery and a small number of genetic loci. However, high-resolution genome monitoring techniques using highly multiplexed FISH probes have recently revolutionized DNA FISH, enabling previously unheard-of scale distance estimation among millions of chromosomal spots in individual cells. In line with previous low-throughput observations (Nora et al., 2012), DNA FISH exposes subpopulation configurations that are unable to be deconvolved from population-averaged Hi-C data, and it quantifies exact separations within the cytoplasm. Furthermore, uncertainties regarding the actual distance associated with Hi-C contact frequency can be resolved with the use of high-resolution imaging. For instance, while a section is identified by Hi-C as barely connected, it is similarly likely that that area continually occurs at a middle separation from the firmly divided A and B type areas of the genome, or that it alternates between significant connections with A and B in various tissues. The more robust A and B sections of the majority of cells reside among these "weak compartment" portions, as shown by chromosomal tracing, rather than oscillating between the extremes (Wang et al., 2016).

However, in order to provide answers to a range of scientific queries, it is vital to differentiate between concurrent and alternative encounters. Does the combined exclusion of distinct regulators in different populations or the coordinated action of many enhancers control gene expression? Is the average of many different loops represented by layered interactions between many CTCF-bound areas, or several loops brought together cooperatively by these interactions? Assessing concurrent interactions is also required to find out if functional gene clusters in transcribed factories are arranged in groups (Sutherland and Bickmore, 2009).

6.2 MULTI-WAY CHROMOSOME INTERACTIONS

Some techniques have demonstrated both multi-contact arrangements happening in a limited area and shared beneficial contacts between numerous sites that are part of the cell community, which would be absent from demographic-averaged maps of pairs of contacts. Concurrent contact detection in cells additionally showed that in the absence of the cohesion-unloading component, if cohesion is not effectively released from DNA, CTCF-loop anchors clash (Allahyar et al., 2018). In general, multiway connections are more inclined to happen among areas inside TADs, and collaborative relationships among a number of loci located on various chromosomes are extremely rare. As a result, it is unlikely that these collaborative interchromosomal connections are necessary for significant biological functions in the majority of cases (Olivares-Chauvet et al., 2016). Although vicinity ligation-based approaches have provided helpful insights, their capacity to record concurrent events is intrinsically limited. Regardless of whether any additional gene loci were within reach, an inhibitor segment can only connect directly between two additional nucleotides in a single cell. Therefore, even while single-cell Hi-C maps (explained in greater detail in the following section) by definition depict sets of concurrent interactions (Atarashi et al., 2013), no single map can accurately represent all interactions occurring in that cell due to extensive scientific subsampling in the latest generations of single-cell Hi-C techniques. From large collections of this data, algorithmic techniques may be able to identify probable concurrent or unique connections. The technique of principal component analysis (PCA) has been used in the field of protein structure to examine the linked formation and breaking of contacts across a sizable set of interaction grids using fluid peptide shape shots (Leon et al., 2016).

Using this method, it has been possible to identify the types of sector associations that most effectively differentiate hemoglobin type chromosomal bending arrangements. It has also shown certain sector connections that have a tendency to establish and split among individual cells in a group of cells. In the near future, this method may be used to analyze single-cell Hi-C contact maps to infer collaborative or antagonistic sets of interactions, even in situations where interactions cannot all be recorded simultaneously in just one cell. The creation of supplementary techniques has also been influenced by the shortcomings of 3C-based techniques for capturing multi-contact exchanges, such as the SPRITE method outlined earlier (Quinodoz et al., 2018). Due to the lack of ligation, it is possible to detect many sites working simultaneously within a single region. A Hi-C map's mean substantial contact rates among all the A compartment areas are the consequence of long-distance interactions inside the A compartment, which tend to take place within an identical cell, but other A segment sections touch one another sporadically, according to SPRITE data (Beagrie et al., 2017). Further records intricate multi-way connections and have found many three-way relationships among TADs with super-enhancers or TADs with strongly expressed areas located in single cells where cooperative, multidirectional interactions take place concurrently (Bintu et al., 2018).

Finally, substantial evidence for cooperative, coordinated interactions among the nuclear membrane and chromatin in particular cells has been found through the single-cell utilization of DamID. It is important to highlight that neither of these techniques can conclusively establish that multiple interactions are necessary for loci to communicate rules information. To find out how they work and develop, as will be clarified below, disturbances of certain mixtures of relationships will be required (Kant et al., 2015).

6.3 CHROMOSOME STRUCTURE IN SINGLE CELLS

3C data typically reflect images of chromosomal configurations at an identifiable point, aggregated over a whole cell population, including the possible exception of just one cell Hi-C as well as certain multi-contact techniques. However, how exactly do they connect to and change over time in relation to the real histone strand configurations in individual cells? Prior to the development of high-resolution chromosome mapping, studies using DNA FISH and live-cell imaging in yeast, *Drosophila*, and human cells demonstrated that genomic orientation varies greatly within groups of cells or tissues (Jhunjhunwala et al., 2008). Only after summing across several individual cells can interaction patterns like TADs and divisions shown in crowd Hi-C maps become apparent, suggesting that they represent favored relationships in a very varied array of components. Because individual-cell Hi-C presently has a restricted genomic conclusion, it is not possible to statistically evaluate immediate contacts like single enhancer/promoter interactions, which are usually less than 200 kb within single TADs. Super-resolution DNA FISH and polymer models have demonstrated the significant variability of specific forms of the nucleosome thread within single TADs; nevertheless, the fact that only a portion of interactions truly take place in a single cell indicates that enhancer-promoter connections and CTCF loops are random processes (Giorgetti et al., 2014) (Figure 6.1).

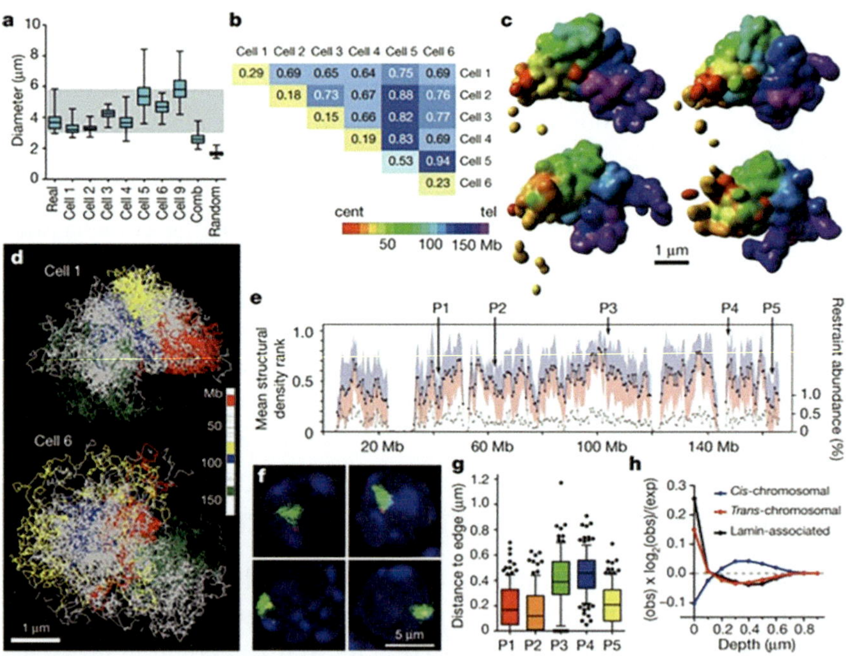

FIGURE 6.1 Single-cell Hi-C demonstrates variation in chromosomal structure from cell to cell.

Source: Reprinted with permission from Nagano et al., 2013. Copyright © 2013, Springer Nature Limited

Stochastic connections inside TADs are preferred over contacts across borders, and TAD borders themselves result from the stacking of shapes when cohesion is lacking. Preferred connections within TADs are lost, and interaction unpredictability increases, which is most likely the cause of the elimination of TAD borders from the population Hi-C matrix (Bintu et al., 2018). This supports the theory that connections inside TADs in each cell represent the recorded location of loop ejection combinations, reflecting a caught moment of an ongoing procedure rather than static loops (Hansen et al., 2018). Additionally, single-cell DamID studies revealed significant inter-cell heterogeneity in LAD interactions with the nuclear periphery (Kant et al., 2015). Thus, significant structural diversity is supported by experimental data throughout all genomic size scales, starting with sub-TAD structures (Boettiger et al., 2016). Given that probabilistic reactions among regulatory chains are anticipated to occur in the probabilistic transmission of regulatory information, ubiquitous cell-to-cell architectural heterogeneity may have significant effects on transcriptional control. Further discussion of techniques that will aid in defining the functional significance of structural variation may be found in Giorgetti et al. (2016).

6.4 CHROMOSOME STRUCTURE IN TIME

The durations during which cell-to-cell structural variations are formed and their potential connections to induction and other nuclear activities are not well understood. In a highly speculative case, a morphological position acquired at mitotic exit is sustained over an entire cell cycle and is partially shuffled and transferred to dividing cells. Under this situation, legislative connections are steady across the course of the cell phase, with some interactions lasting for longer periods and others never happening. The opposite scenario would be for each cell to go through a great number of distinct shapes in its own cycle. In these circumstances, the sharing of regulatory data would be an extremely erratic and volatile process, with links coming together and breaking apart in a matter of minutes or seconds. Taking into account the fact that the amount of time that should elapse for two loci on a chromosome to meet again is predicted to rise quickly in proportion to their genomic distance (Zhang and Dudko, 2016). In actuality, both possibilities are possible. In contrast to the possibility of slower changes in location among motifs placed hundreds of kilobases apart, links among regulatory regions separated by short genomic lengths may build and deconstruct quickly (Gibcus and Dekker, 2013). Despite identifying particular patterns, a number of methods have made it possible to image chromosomal oscillations at the size of the entire nucleus. Using photoactivatable fluorescence for identifying histone proteins has demonstrated that chromosomal territories remain in the same locations during interphase (Strickfaden et al., 2010). Comparable findings were reached where areas linked to the Dam-Lamin B1 crossover protein are marked by a luminous marker of N6-methylated adenines. Granular stream restoration and association have been used to study shorter-scale intra-chromosomal changes (Mohamed et al., 2018).

Using a variety of complementary proteins that are fluorescent, it is necessary to concurrently monitor multiple regions of the genome in living cells in order to measure periodic changes in chromosomal arrangement in a sequence-specific manner. By inserting collections of irregular transcription element receptors, such as Lac or Tet operators, through homologous recombination, investigators have traditionally been able to observe specific chromosomal loci. These spots can be restricted to diffraction-limited areas once the appropriate inhibitors combined with luminescent proteins interact with them (Kumaran and Spector, 2008). It is necessary to target consecutively and laboriously to find genomically viable subgroups when inserting multiple parallel arrays in cis. It is essential to express many gRNA arrays and several inactive catalytic Cas proteins simultaneously in order to use multi-color CRISPR techniques. Furthermore, lenses that can capture signals of excellent quality from dim sub-diffraction illumination while delivering low photon counts (to limit photobleaching and phototoxicity) are required for observing cells in biological settings across durations spanning from seconds to hours. These restrictions have been addressed in two recent investigations that used orthogonal operator arrays to examine the temporal progression of two genomic regions in cis in mouse cells. The mean location of the regions bordering the Sox2 gene and its (super-) enhancer, which are situated around 100 kb apart, varies in every single cell of embryonic stem cells from mice (Alexander et al., 2019).

Although the subjects' immediate locations varied considerably from the mean throughout the experiment, the median placements were constant over a 25-minute trial time window, with occasional sharp fluctuations. The scenario is analogous in B cell precursors, since it was discovered that the temporal placements of the VH and DHJH sections at the antibody loci (divided by approximately 1 Mb) vary around constant typical values on a 10-minute timeline, which can shift across more time in a particular group of lymphocytes (Khanna et al., 2019). Although the subjects' immediate locations varied considerably from these mean values as time passed, overall placements were constant over a 25-minute testing period, with occasional sharp variations. Similar circumstances were discovered in B cell precursors, when it was shown that the relative positions of the VH and DHJH sections at the antibody loci (divided by approximately 1 Mb) fluctuated around constant medians on a 10-minute timeline, which can alter over greater periods in certain groups of cells (Khanna et al., 2019).

Up to about 100 kb/min, cohesion is capable of extruding DNA rings. Therefore, it ought to be expected that the circular insertion operation influences the motion of loci isolated by several kilobases across minutes-long timespans. Alternatively, translucency-independent monomer activity might be represented by shorter-timescale (<1 min) oscillations surrounding steady locations. In general, our findings imply that enhancer-promoter connections may manifest as erratic and fleeting random occurrences on multiple occasions during a cell cycle. Different research observed the relationship between the indigenous dawn loci, which are situated 100 kb away from the exogenous even skipped (eve) booster, in live *Drosophila* embryos (Chen et al., 2018). An aberrant isolator that can mate with its indigenous counterpart at the eve region was used to impose cycling across the two loci. Remarkably, compared to studies in mouse cells, the actual contacts among the two loci continued for a couple of minutes (Alexander et al., 2019), and the looping condition as well as translation on the reporter gene's location served to further consolidate them.

The ability to expand and extend these findings to different species and circumstances is becoming more possible due to developments in molecular programming and imaging. In addition to CRISPR-based knock-in techniques, activator array implantation may be effectively achieved by transposon-based and recombinase-mediated genetic editing (Redolfi et al., 2019). This makes it possible to assess real specimens in technicolor with great precision and little toxicity. Alongside developments in tagging techniques and super-resolution live-cell monitoring, such tools will enable investigators to explore basic issues like whether and how topological limitations, loop extensions, and CTCF loops affect genome interaction interactions and how structural variation connects to the fluid trade of legislative data that ultimately impacts the transcription process (Zheng and Lavis, 2017).

6.5 POLYMER MODELS AND CHROMOSOME STRUCTURE

Translation of interactions into measurements, concurrent interaction inference, and representation of variation in structures among tissues and over time are just a few of the difficulties in interpreting chromosome data regarding structure that mathematical simulation of 3C data and chromosome structure offers as an important way to address. Structural histone simulations have been shown to be effective tools for analyzing the spatial structure of the genome for a variety of reasons. Firstly, the fundamental framework must be inferred from mathematical representations, as both optical and 3C-based techniques do not yet permit an immediate, excellent quality, massive observation of the structure of chromatin. Furthermore, models offer a numerical structure for combining data from many practical system types. Finally, modeling methodologies frequently enable us to immediately investigate theories about fundamental chemical reactions and to provide empirically testable conclusions. The most evident objective of structural chromatin models is to identify the underlying structures, as 3C-based techniques evaluate association probabilities instead of creating real formations. To look for a single structure is the easiest strategy (Duan et al., 2010).

This procedure assumes two crucial things: that the entire range of components detected by bulk genomic approaches may be represented by a single structure, and that it is possible to translate connection probabilities into spatial distances. Because genomic structure varies greatly, the second assumption presents challenges, even though the first premise may be valid (as previously explained). Owing to the drawbacks of employing a single genomic framework, modeling techniques have been expanded to find a collection of shapes whose averages coincide with observed population-average interaction rates (Pascual et al., 2017). Even if this is more in line with physiological reality, a new difficulty arises. The issue is severely under-constrained since these approaches attempt to reconstruct a distribution by taking into account just its average, even though other sets of structural variations can be compatible with observed population-averaged association patterns. Simulations using molecular dynamics and energy landscapes are two methods used to limit the set of solutions (Di Pierro et al., 2016).

It is essential that these techniques produce feasible hypotheses that may be independently verified alongside confirmation through *in silico* simulations. For instance, the variation of lengths between certain pairs of loci within the population may be tested using microscopy, and the results can be matched to the pattern of lengths generated by models (Giorgetti et al., 2014). It's still uncertain if this kind of technique yields quantitative projections and precise depictions of genomic structure, even with empirical confirmations. Single-cell Hi-C investigations circumvent the issue of population aggregation constraints; however, the precision of models derived from single-cell Hi-C maps is severely constrained by their poor yield per cell. A different category of modeling techniques begins with basic physical or chemical operations and works to create structures that make sense of data on interactions based on three dimensions. The advantage of these techniques is that they may test theories about cellular functions and the constantly changing events underpinning chromosome shape directly, even if they often yield less precise data reproduction. Various biophysical models have been suggested to elucidate characteristics of high-temperature data: According to the spiral globule hypothesis, communication probability in human cells is inversely correlated with genetic distance (Lieberman-Aiden et al., 2009).

One notable computational accomplishment is the loop ejection model, which offers a condensed method that clarifies both the shape of mitochondrial chromosomes and the development of TADs (Fudenberg et al., 2016). The model generated feasible hypotheses, which were confirmed in genomic alterations including WAPL modifications, NIPBL, and CTCF (Haarhuis and Rowland, 2017). Nevertheless, there are still several aspects of the model that cannot be fully understood *in vivo*, including the chemical mechanism underpinning shielding and the pace at which extrusion occurs—both uni- and simultaneous. Combining the processes underlying interaction patterns at various sizes will be an exceptional task (Nuebler et al., 2018). Potential pairing of phase division and loop protrusion to produce distance, TAD, and structures for loops. Furthermore, new techniques will be needed to methodically integrate 3C data with microscope data (Abbas et al., 2019). Ultimately, examining the aspects of the data that systems are unable or unwilling to adequately describe will be just as helpful as emphasizing the elements that the models are able to adequately explain.

6.6 CONNECTING CHROMOSOME STRUCTURE TO FUNCTION

Accurate assessments of the connections among genetic loci are unable to identify which relationships are associated with (or cause) functional processes, even whether they are converted into complex polymer models or validated by high-resolution imaging. It is assumed that interaction rates between a number of loci in the genome provide information regarding the likelihood of functional communication between those loci. The development of computer techniques to find "specific," significant, or "looping" contacts in Hi-C or 4C data has therefore become a trend in the area. These are typically described as pairings of repeats whose interaction patterns jump away from the relationship backdrop created by stochastic "collisions" throughout the DNA polymer; however, the terminology differs significantly between research relationships with a range of analytical techniques. The outcomes of these methods can differ greatly, ranging from finding enhanced spot contacts (Rao et al., 2014). Furthermore, the specifics of the research methodology may have some impact on the apparent characteristics of these relationships in an interaction map. In contrast to previous Hi-C datasets, which show more uniform lines, 3C approaches that shorten DNA strands prior to binding, such as micro-C, identify grids of particular contact dots (Hsieh et al., 2019).

It is difficult to choose the optimal looping process or domain-calling method (or the optimal looping process or domain-resolving practical approach) in the absence of a thorough understanding of the physiological implications of various kinds of "significant" interactions. Above-background events are unquestionably important in comparison to a null model in which collisions are purely random. They closely associate with CTCF loops, gene-active gene connections, or gene-potential legislative region connections (Bonev et al., 2017). Nonetheless, a sizable portion of regulatory chains does not exhibit connections to adjacent mates that are very rich. Does this mean that these relationships aren't necessarily functional? The average increase in saturation levels at "specific" encounters is only 0.5 times greater than the background of random collisions, with maximum increases of 2 to 3 times. Can other biological processes, such as transcribed control, take into account even minute variations in interaction probability? On the other hand, should low-frequency connections at short chromosomal regions that are not considered important but yet occur as ambient protein impacts be discounted as inactive? Connections across genetic sequences that are widely apart, either in trans (i.e., on separate chromosomes) or in cis (1 Mb on the same chromosome), may be subject to similar issues because of the nearly random organization of chromosomal areas in trans and the power-law decline of contact probability in cis. Such interactions are not represented in 3C ligase collections because they are inherently considerably uncommon compared to those between sequences that are close together in cis (<1 Mb). Since they're so uncommon far-cis (Van De Werken et al., 2012).

A noteworthy finding from this work is that long-range cis and trans connections have emerged in sensory nerve cells, directly influencing how they operate through the choice of specific receptors for smell (Monahan et al., 2019). This raises the question of whether solitary connections may be functionally significant, or if they might instead represent (perhaps cooperative) synergy between multiple combinations that occur in various cells. The development of testing techniques that employ molecular readouts to identify "productive" functional connections rather than just identifying contacts that are noticeably enhanced is necessary to address these issues. Circumstantial disruptions, like the reduction of CTCF and cohesion, have indicated that, in general, the "important" structures linked to CTCF/cohesion are crucial for the long-term regulation of a majority of genes (Rao et al., 2017). Nonetheless, it remains challenging to determine the local causation and significance of individual interactions while numerous events are impacted simultaneously. Reprogramming of genomic interactions is caused by specific chromosomal changes that alter TAD borders or CTCF-CTCF contacts. These changes have been demonstrated to cause gene deregulations or reduced chronological resilience (Despang et al., 2019).

The intricate impact of histone looping on expression at certain loci has been better understood because of these generated loops. Enforcer-gene promoter interactions were shown to stimulate the transcription of genes, but only within cell types that are generally receptive to the investigated gene's activation, providing evidence that chromosomal interactions can control gene expression (Morgan et al., 2017). It has demonstrated little impact on transcription; it has also been possible to examine whether the stimulated relocation of chromosomal loci to different nuclear divisions affects gene expression using other techniques like CRISPR-GO (Wang et al., 2018).

6.7 CHROMOSOME STRUCTURE AND TRANSCRIPTION

Changes in the interactions between regulatory parts of genomes are hypothesized to modify the function of transcription and/or resilience in gene expression: Greater geographical closeness between the target genes is often correlated with enhancer activity, which is measured by means of chromatin accessibility and/or richness in functional histone modifications (Carter, 2002). TADs are shared by genomic areas where randomly integrated transgenic mice in rodents express themselves in corresponding tissue-specific ways (Symmons et al., 2014). Cancer-causing gene stimulation may arise from aberrant enhancer-promoter interactions caused by elimination, tilting, or duplication of TAD borders (and, to a lesser degree, single CTCF sites) (Hnisz et al., 2016). Nevertheless, the molecular concepts that may connect transcriptional control and chromosomal connections remain mostly unknown. There is growing evidence that modifiers have the ability to control the rate of activator release (Senecal et al., 2014). Without knowing if translation happens concurrently with, after, or apart from physical interactions, it is unclear how this relates to in-person interactions. Studies using live-cell imaging that seek to link the geographical locations of enhancers and promoters to the amounts of RNA produced in individual cells (Chao et al., 2008).

Therefore, the looping condition is stabilized. Nonetheless, in mouse stem cell embryos, there does not seem to be a temporal correlation between transcription levels and the Sox2 gene locus's physical closeness to the area that surrounds its mega-enhancer (Alexander et al., 2019). Surprisingly, insulator-induced loops are uncommon and persistent in *Drosophila*, which may make it easier to find transcriptional connections among physical closeness and activation. In comparison, there are numerous and fleeting contacts at the Sox2 locus. Therefore, it will be fascinating to see if artificially created persistent relationships in mammals correlate more strongly with single-cell transcription states. Nevertheless, it should be mentioned that bacterial operator arrays are used in order to prevent meddling with transcriptional regulation (Alexander et al., 2019).

A small number of kilobases were added outside the actual Sox2 and super-enhancer areas. Due to the possibility of significant chromatin fiber bending in the 2–10 kb range, their spatial placements may not accurately reflect the positions of the regulatory sequences (Redolfi et al., 2019).

There is considerable confusion regarding the geographical locations of the reporter and endogenous genes since these signals are produced by MS2- and PP7-labeled RNA signals. An alternate and well-respected theory on how boosters could communicate legislative data with promoters is that there's no need for a rigid short-range (nanometer-scale) interaction. Under this scenario, transcription element condensation extending dozens of microns might be formed by regulatory sequences, even if they are not in direct contact. It is suggested that low-complexity areas found on various transcription factors influence non-specific interactions to cause this. Recent observations have shown nuclear phase-separated melts (Boija et al., 2018). They are hypothesized to induce transcriptional stimulation by substantially raising transcription factor concentrations in the direct geographical vicinity of promoters. This thought-provoking idea, which draws comparisons to the "transcription factory" paradigm that was previously put forward (Sutherland and Bickmore, 2009), might perhaps offer a reason for the absence of association seen among element length and Sox2 gene transcriptional activity. Additional communication channels have also been suggested, such as "tracking" processes, which transfer regulatory data by 1D transcription factor diffusion along the chromatin fiber. These channels may be particularly important for promoter-proximal enhancers (Vernimmen and Bickmore, 2015).

In a perfect world, promoter regulatory output would be precisely measured and correlated with its 3D distance to promoters split by varying chromosomal lengths and across morphological boundaries in order to elucidate the mechanisms that control regulatory communication mediated by 3D chromosomal structure. Nevertheless, considering the quantity of genomic modifications necessary and the intricate structure and regulation of transcribed genomic areas in eukaryotic genomes, this endeavor continues to be extremely difficult. Studies utilizing genome engineering in *E. coli* are not the primary rate-limiting step; rather, studies have shown that transcription induced by bacterial elements resembling enhancers is closely correlated with the likelihood that these elements will make contact with a promoter and falls off as genomic distance increases like a power law (Hao et al., 2019).

The LacI tetramerization-induced designed loops have the ability to either inhibit or shield production based on their relative location to the "enhancer"-promoter pair. Comparable studies conducted in higher eukaryotes will be fascinating to watch. Lastly, not all genomic areas may have a strong correlation between the transcriptional stage and resilience of a regulator and the frequency of interactions between elements and promoters. The nuclear location of a gene and its closeness to repressive compartments like the nuclear lamina are two important variables that affect gene activity (Akhtar et al., 2013). It's now feasible to construct studies that separate the different transcriptional dependencies on nucleotide structure, nuclear translation, and chromosomal shape, thanks to developments in genetic technology, stimulated protein oxidation structures, scans, and sequencing-based methodologies. A development in recent methods allows transcriptional and nucleosome indicators, or chromosomal localization, to be measured simultaneously in single cells (Rooijers et al., 2019). This may one day enable the monitoring of chromosomal connections inside a single cell, as well as translation or transcription factor binding. These techniques will deepen our knowledge of how molecular processes spanning a variety of timeframes, such as transcription factor binding and loop ejection motion, are combined to produce specific transcriptional outputs, in conjunction with single-molecule investigations in living cells.

6.8 CHROMOSOME STRUCTURE AND OTHER NUCLEAR PROCESSES

The discussion around the effects of genomic bending on transcriptional regulation frequently centers on chromosomal architecture, although other nuclear processes, such as replication timing, DNA damage, repair, and rearrangement creation, may also be significantly impacted by chromosome layout. Multiple strategies exist in which 3D genome folding affects DNA damage and repair. The average 3D proximity between chromosomal regions in cell populations and chromosomal frequency have been found to be correlated, according to data from FISH, rearrangement DNA sequencing, and Hi-C (Parada et al., 2004). Fluorescence-labeled DNA breaks in living cells have been tracked using imaging, and the results have shown that although massive migrations of double-strand breaks (DSBs) occasionally happen, DSB pairing more frequently occurs among breaks that were initially proximal (Roukos et al., 2013). Chromosome shape has an impact on these differences in the development of DNA repair at other locations as well: DSBs that arise in distinct nuclear sites and in heterochromatin, as opposed to euchromatin, display varying degrees of mobility and select distinct repair pathways (Lemaître and Soutoglou, 2014). This spatial clustering of DSBs is demonstrated by chromosomal packing utilizing Capture-C after chromatin abnormalities are induced (Aymard et al., 2017).

The technique of causing and tracking certain heterochromatin breaks has proven difficult thus far. The processes involved in chromosomal folding and DNA repair remarkably share key protein components. Although CTCF and cohesion have long been known to react to DNA damage, they are now emerging as important components of genomic architecture (Birkenbihl and Subramani, 1992). More links between these parameters related to genome architecture and damage to DNA have been discovered as a result of novel experimental methodologies. The CTCF TAD border sites are susceptible to breaking because of topological strain, according to accurate identification of DNA DSBs using END-seq (Canela et al., 2017). Additionally, the method for DNA healing may have significant effects on chromosomal organization. Enhancing agents and promoters that are typically isolated could get juxtaposed if DNA damage repair breaks a TAD border, which might lead to the errors of cancer genes (Hnisz et al., 2016).

Deoxyribonucleic acid damage repair may preserve the integrity of the 3D genomic structure through the interactions between chromosomal architectural proteins and DNA repair factors. In fact, extra insulation between TADs linked to DNA damage repair is present across a cell population, as determined by Hi-C (Sanders et al., 2019). By doing this, misrepair across TAD bounds may be avoided. Investigations into possible links between DNA damage repair, TADs, and loop extrusion will require more effort. While it is sometimes necessary to insert synthetic sequences into the genome in order to monitor individual DSBs, population measures are unable to track genome interactions and damage to DNA at the same locus in the same cell. A more accurate evaluation of the relationships between DNA DSBs and alterations in genome structure might be possible with linked single-cell profiling using multiple genomic approaches integrated together. Single-cell multi-omics has so far assessed RNA-seq and ATAC-seq in the same cells (Cao et al., 2018). Methods like END-seq may eventually be added to this idea. After all, early duplicating areas are found in the A compartment, and late copying regions are found in the B segment. Replication time (RT) and A/B segment identities correlated highly. Likewise, RT shifts between cell types include modifications to whole TADs, and RT domain borders correspond with TAD boundaries (Pope et al., 2014). Compared with distinct chromatin changes, these RT regions in differentiated cell types have a stronger correlation with the A/B division organization (Dixon et al., 2012).

For this reason, two complementary methods that both highlight the scientific significance of the same kind of region organization are chromosomal contact measurements and RT. Local RT shifts have been found by replication timing patterns between patients and controls, which may provide insight into disease indicators. Thus, RT profiling can be an effective, high-quality method to read out any variations in local chromosomal condition and shape because of the significant association that exists between duplication time and chromosome structure (Rivera-Mulia et al., 2017). Further investigation is required to elucidate the consequences and mechanisms of chromosomal structure creation that are shown by the association among RT and geographic divisions. According to recent research, the deletion of local border regions does not disrupt either RT domains or A/B compartments. However, the deletion of specific sites within domains known to be initial replication control elements (ERCEs) does disrupt both RT and compartments (Sima et al., 2019). By influencing the present-day chromatin patterns and gene regulation in the surrounding area, these areas seem to have an impact on both RT and compartmentalization. On the other hand, RT variations that do not correspond with A/B compartment alterations occur during very early stages of cell fate selection and division (Dileep and Pournami, 2019). While physical interaction alterations and RT shifts are often tightly connected, further research is required to understand how histone markers and other variables can occasionally define RT changes without these modifications. Chromosomes are large, physical structures that affect a cell's physical features along with storing, replicating, and transmitting gene data (Bustin and Misteli, 2016).

Methods for manipulating cells and photographing them are illuminating potential effects of chromosomal arrangement on the fundamental properties of cells. Chromosome status affects nucleus flexibility, as demonstrated by stretching particular nuclei: more heterochromatin stiffens nuclei, whereas a protein called increases nuclei's suppleness (Banigan et al., 2017). On the other hand, physical factors can modify the structure of the genome: these protein regions are occasionally irreversibly stretched in cores inhaled via thin micropipettes (Irianto et al., 2017). These findings imply that distinct chromosomal architecture may be more important for preserving nuclear mechanical integrity than for regulating particular genes, or that physical and regulatory functions may be linked. Measurements of single-cell mechanical characteristics and nucleus deformations require new methods and more effort to close the distance that separates the particular to the sequence contact details of chromosomal orientation acquisition methods. Large-scale physical manipulation of cells or nuclei is becoming achievable, and thus opens the possibility of connecting these alterations to bulk genomic data (Earle et al., 2019). However, further investigation is required to tackle the variability of initial chromosomal configurations in each cell and their impact on the genome's response to physical disruption.

It makes sense that the discussion of chromosomal design has moved from how to quantify 3D structures with greater resolution and realism to how to comprehend the forces that shape these frameworks and how to assess their usefulness within the larger framework of nuclear activities. A growing number of researchers are using 3C methodologies, including Hi-C in particular, thus it will be critical to consider what qualities these various approaches can represent. As this chapter discusses, population-averaged 3C-based measures cannot capture all of the variation that exists in space; therefore, some care is necessary. While 3D distances may be adequately represented by signals in 3C approaches, the opposite is also true. Furthermore, an elevated degree of trust regarding interactions above ambient polymerization activity may not always imply a correlation with physiological activity in paired contacts evaluated by 3C. However, recent advances in alternative approaches have demonstrated that the population-averaged 3C techniques reflect genuine characteristics of chromosomal bending, given these theoretical and technological restrictions. The ability to statistically analyze 3C data in regard to interaction chances is noteworthy because it opens the door to a fully scientific understanding of chromosomal architecture and the processes that shape it. Methods that integrate focused disruption techniques, fundamental theoretical frameworks, and more sophisticated and quantified 3C methodologies are anticipated to provide significant advancements in this field.

Interestingly, the majority of statistically significant traits discovered in 3C data can sometimes differ from the most physiologically significant ones, as the hunt for connections between chromosomal shape and physiological activity intensifies. For instance, transcriptional alterations may be triggered by transient conversations as opposed to just secure conversations, which makes relationships with structural modifications difficult to find, or by interactions that are more distant and controlled by microphage-separated condensation as opposed to locus-specific localization incidents. The next step in describing the effect of chromosomal interactions on operation is likely to require models combining intricate delicate and fluid modifications into general shifts in phenotype, much as systems biologists have progressed from simple predictable pathways of signaling to novel characteristics of intricate legislative systems. Developing live-cell imaging methods that are more advanced will undoubtedly be crucial in this regard. However, novel methods like cryo-electron tomography, which let scientists see macromolecular assemblies within cells in real time, may potentially make discoveries possible (Bäuerlein et al., 2017).

KEYWORDS

- **3D genome organization**
- **chromosomal architecture**
- **chromosome rearrangements**
- **DNA double-strand breaks**
- **methodologies**
- **replication timing**
- **topologically associating domains**

REFERENCES

Abbas, J., Aman, J., Nurunnabi, M., & Bano, S. (2019). The impact of social media on learning behavior for sustainable education: Evidence of students from selected universities in Pakistan. *Sustainability, 11*(6), 1683.

Akhtar, M. S., Panwar, J., & Yun, Y. S. (2013). Biogenic synthesis of metallic nanoparticles by plant extracts. *ACS Sustainable Chemistry & Engineering, 1*(6), 591–602.

Alexander, B., Ashford-Rowe, K., Barajas-Murph, N., Dobbin, G., Knott, J., McCormack, M., Pomerantz, J., Seilhamer, R., & Weber, N. (2019). *Horizon Report 2019 Higher Education Edition.* EDU19.

Allashyar, A., Vermeulen, C., Bouwman, B. A., Krijger, P. H., Verstegen, M. J., Geeven, G., van Kranenburg, M., Pieterse, M., Straver, R., & Haarhuis, J. H. (2018). Enhancer hubs and loop collisions identified from single-allele topologies. *Nature Genetics, 50*(8), 1151–1160.

Atarashi, K., Tanoue, T., Oshima, K., Suda, W., Nagano, Y., Nishikawa, H., Fukuda, S., Saito, T., Narushima, S., & Hase, K. (2013). Treg induction by a rationally selected mixture of *Clostridia* strains from the human microbiota. *Nature, 500*(7461), 232–236.

Aymard, F., Aguirrebengoa, M., Guillou, E., Javierre, B. M., Bugler, B., Arnould, C., Rocher, V., Iacovoni, J. S., Biernacka, A., & Skrzypczak, M. (2017). Genome-wide mapping of long-range contacts unveils clustering of DNA double-strand breaks at damaged active genes. *Nature Structural & Molecular Biology, 24*(4), 353–361.

Banigan, E. J., Stephens, A. D., & Marko, J. F. (2017). Mechanics and buckling of biopolymeric shells and cell nuclei. *Biophysical Journal, 113*(8), 1654–1663.

Bäuerlein, F. J., Saha, I., Mishra, A., Kalemanov, M., Martínez-Sánchez, A., Klein, R., Dudanova, I., Hipp, M. S., Hartl, F. U., & Baumeister, W. (2017). *In situ* architecture and cellular interactions of PolyQ inclusions. *Cell, 171*(1), 179–187.e110.

Beagrie, R. A., Scialdone, A., Schueler, M., Kraemer, D. C., Chotalia, M., Xie, S. Q., Barbieri, M., de Santiago, I., Lavitas, L. M., & Branco, M. R. (2017). Complex multi-enhancer contacts captured by genome architecture mapping. *Nature, 543*(7646), 519–524.

Belmont, P. J., Goodman, G. P., Waterman, B. R., Bader, J. O., & Schoenfeld, A. J. (2014). Thirty-day postoperative complications and mortality following total knee arthroplasty: Incidence and risk factors among a national sample of 15,321 patients. *JBJS, 96*(1), 20–26.

Bintu, B., Mateo, L. J., Su, J. H., Sinnott-Armstrong, N. A., Parker, M., Kinrot, S., Yamaya, K., Boettiger, A. N., & Zhuang, X. (2018). Super-resolution chromatin tracing reveals domains and cooperative interactions in single cells. *Science, 362*(6413), eaau1783.

Birkenbihl, R. P., & Subramani, S. (1992). Cloning and characterization of *rad21*, an essential gene of *Schizosaccharomyces pombe* involved in DNA double-strand-break repair. *Nucleic Acids Research, 20*(24), 6605–6611.

Boettiger, A. N., Bintu, B., Moffitt, J. R., Wang, S., Beliveau, B. J., Fudenberg, G., Imakaev, M., Mirny, L. A., Wu, C. T., & Zhuang, X. (2016). Super-resolution imaging reveals distinct chromatin folding for different epigenetic states. *Nature, 529*(7586), 418–422.

Boija, A., Klein, I. A., Sabari, B. R., Dall'Agnese, A., Coffey, E. L., Zamudio, A. V., Li, C. H., Shrinivas, K., Manteiga, J. C., & Hannett, N. M. (2018). Transcription factors activate genes through the phase-separation capacity of their activation domains. *Cell, 175*(7), 1842–1855.e1816.

Bonev, B., Cohen, N. M., Szabo, Q., Fritsch, L., Papadopoulos, G. L., Lubling, Y., Xu, X., Lv, X., Hugnot, J. P., & Tanay, A. (2017). Multiscale 3D genome rewiring during mouse neural development. *Cell, 171*(3), 557–572.e524.

Bustin, M., & Misteli, T. (2016). Nongenetic functions of the genome. *Science, 352*(6286), aad6933.

Canela, A., Maman, Y., Jung, S., Wong, N., Callen, E., Day, A., Kieffer-Kwon, K. R., Pekowska, A., Zhang, H., & Rao, S. S. (2017). Genome organization drives chromosome fragility. *Cell, 170*(3), 507–521.e518.

Cao, Y., Fatemi, V., Fang, S., Watanabe, K., Taniguchi, T., Kaxiras, E., & Jarillo-Herrero, P. (2018). Unconventional superconductivity in magic-angle graphene superlattices. *Nature, 556*(7699), 43–50.

Carter, M. R. (2002). Soil quality for sustainable land management: Organic matter and aggregation interactions that maintain soil functions. *Agronomy Journal, 94*(1), 38–47.

Chao, B. F., Wu, Y. H., & Li, Y. (2008). Impact of artificial reservoir water impoundment on global sea level. *Science, 320*(5873), 212–214.

Cleynen, I., Boucher, G., Jostins, L., Schumm, L. P., Zeissig, S., Ahmad, T., Andersen, V., Andrews, J. M., Annese, V., & Brand, S. (2016). Inherited determinants of Crohn's disease and ulcerative colitis phenotypes: A genetic association study. *The Lancet, 387*(10014), 156–167.

Despang, A., Schöpflin, R., Franke, M., Ali, S., Jerković, I., Paliou, C., Chan, W. L., Timmermann, B., Wittler, L., & Vingron, M. (2019). Functional dissection of the Sox9–Kcnj2 locus identifies nonessential and instructive roles of TAD architecture. *Nature Genetics, 51*(8), 1263–1271.

Di Pierro, M., Zhang, B., Aiden, E. L., Wolynes, P. G., & Onuchic, J. N. (2016). Transferable model for chromosome architecture. *Proceedings of the National Academy of Sciences, 113*(43), 12168–12173.

Dileep, M., & Pournami, P. (2019). AyurLeaf: A deep learning approach for classification of medicinal plants. *TENCON 2019–2019 IEEE Region 10 Conference (TENCON).*

Dixon, J., Levine, M., Reicher, S., & Durrheim, K. (2012). Beyond prejudice: Are negative evaluations the problem and is getting us to like one another more the solution? *Behavioral and Brain Sciences, 35*(6), 411–425.

Duan, Z., Andronescu, M., Schutz, K., McIlwain, S., Kim, Y. J., Lee, C., Shendure, J., Fields, S., Blau, C. A., & Noble, W. S. (2010). A three-dimensional model of the yeast genome. *Nature, 465*(7296), 363–367.

Earle, M., Hodson, G., Dhont, K., & MacInnis, C. (2019). Eating with our eyes (closed): Effects of visually associating animals with meat on antivegan/vegetarian attitudes and meat consumption willingness. *Group Processes & Intergroup Relations, 22*(6), 818–835.

Franke, M., Ibrahim, D. M., Andrey, G., Schwarzer, W., Heinrich, V., Schöpflin, R., Kraft, K., Kempfer, R., Jerković, I., & Chan, W. L. (2016). Formation of new chromatin domains determines pathogenicity of genomic duplications. *Nature, 538*(7624), 265–269.

Fudenberg, G., Abdennur, N., Imakaev, M., Goloborodko, A., & Mirny, L. A. (2017). Emerging evidence of chromosome folding by loop extrusion. *Cold Spring Harbor Symposia on Quantitative Biology, 82,* 45–55.

Fudenberg, G., Imakaev, M., Lu, C., Goloborodko, A., Abdennur, N., & Mirny, L. A. (2016). Formation of chromosomal domains by loop extrusion. *Cell Reports, 15*(9), 2038–2049.

Gibcus, J. H., & Dekker, J. (2013). The hierarchy of the 3D genome. *Molecular Cell, 49*(5), 773–782.

Giorgetti, L., Galupa, R., Nora, E. P., Piolot, T., Lam, F., Dekker, J., Tiana, G., & Heard, E. (2014). Predictive polymer modeling reveals coupled fluctuations in chromosome conformation and transcription. *Cell, 157*(4), 950–963.

Giorgetti, L., Lajoie, B. R., Carter, A. C., Attia, M., Zhan, Y., Xu, J., Chen, C. J., Kaplan, N., Chang, H. Y., & Heard, E. (2016). Structural organization of the inactive X chromosome in the mouse. *Nature, 535*(7613), 575–579.

Haarhuis, J. H., & Rowland, B. D. (2017). Cohesin: Building loops, but not compartments. *The EMBO Journal, 36*(24), 3549–3551.

Hansen, D. V., Hanson, J. E., & Sheng, M. (2018). Microglia in Alzheimer's disease. *Journal of Cell Biology, 217*(2), 459–472.

Hao, L., Kang, L., Huang, H., Ye, L., Han, K., Yang, S., Yu, H., Batmunkh, M., Zhang, Y., & Ma, T. (2019). Surface-halogenation-induced atomic-site activation and local charge separation for superb CO_2 photoreduction. *Advanced Materials, 31*(25), 1900546.

Hastings, J., Owen, G., Dekker, A., Ennis, M., Kale, N., Muthukrishnan, V., Turner, S., Swainston, N., Mendes, P., & Steinbeck, C. (2016). ChEBI in 2016: Improved services and an expanding collection of metabolites. *Nucleic Acids Research, 44*(D1), D1214–D1219.

Hnisz, D., Day, D. S., & Young, R. A. (2016). Insulated neighborhoods: Structural and functional units of mammalian gene control. *Cell, 167*(5), 1188–1200.

Hsieh, C. T., Hurst, E., Jones, C. I., & Klenow, P. J. (2019). The allocation of talent and U.S. economic growth. *Econometrica, 87*(5), 1439–1474.

Imakaev, M., Fudenberg, G., McCord, R. P., Naumova, N., Goloborodko, A., Lajoie, B. R., Dekker, J., & Mirny, L. A. (2012). Iterative correction of Hi-C data reveals hallmarks of chromosome organization. *Nature Methods, 9*(10), 999–1003.

Irianto, J., Xia, Y., Pfeifer, C. R., Athirasala, A., Ji, J., Alvey, C., Tewari, M., Bennett, R. R., Harding, S. M., & Liu, A. J. (2017). DNA damage follows repair factor depletion and portends genome variation in cancer cells after pore migration. *Current Biology, 27*(2), 210–223.

Jhunjhunwala, S., van Zelm, M. C., Peak, M. M., Cutchin, S., Riblet, R., van Dongen, J. J., Grosveld, F. G., Knoch, T. A., & Murre, C. (2008). The 3D structure of the immunoglobulin heavy-chain locus: Implications for long-range genomic interactions. *Cell, 133*(2), 265–279.

Kant, P., Kind, O., Kintscher, T., Lohse, T., Martini, T., Mölbitz, S., Rieck, P., & Uwer, P. (2015). HatHor for single top-quark production: Updated predictions and uncertainty estimates for single top-quark production in hadronic collisions. *Computer Physics Communications, 191*, 74–89.

Kumaran, R. I., & Spector, D. L. (2008). A genetic locus targeted to the nuclear periphery in living cells maintains its transcriptional competence. *The Journal of Cell Biology, 180*(1), 51–65.

Lemaître, C., & Soutoglou, E. (2014). Double strand break (DSB) repair in heterochromatin and heterochromatin proteins in DSB repair. *DNA Repair, 19*, 163–168.

Leon, M. B., Smith, C. R., Mack, M. J., Makkar, R. R., Svensson, L. G., Kodali, S. K., Thourani, V. H., Tuzcu, E. M., Miller, D. C., & Herrmann, H. C. (2016). Transcatheter or surgical aortic-valve replacement in intermediate-risk patients. *New England Journal of Medicine, 374*(17), 1609–1620.

Lieberman-Aiden, E., Van Berkum, N. L., Williams, L., Imakaev, M., Ragoczy, T., Telling, A., Amit, I., Lajoie, B. R., Sabo, P. J., & Dorschner, M. O. (2009). Comprehensive mapping of long-range interactions reveals folding principles of the human genome. *Science, 326*(5950), 289–293.

Mifsud, B., Tavares-Cadete, F., Young, A. N., Sugar, R., Schoenfelder, S., Ferreira, L., Wingett, S. W., Andrews, S., Grey, W., & Ewels, P. A. (2015). Mapping long-range promoter contacts in human cells with high-resolution capture Hi-C. *Nature Genetics, 47*(6), 598–606.

Mohamed, F., Abukhadra, M. R., & Shaban, M. (2018). Removal of safranin dye from water using polypyrrole nanofiber/Zn-Fe layered double hydroxide nanocomposite (Ppy NF/Zn-Fe LDH) of enhanced adsorption and photocatalytic properties. *Science of the Total Environment, 640*, 352–363.

Monahan, K., Horta, A., & Lomvardas, S. (2019). LHX2- and LDB1-mediated trans interactions regulate olfactory receptor choice. *Nature, 565*(7740), 448–453.

Morgan, P. L., Farkas, G., Hillemeier, M. M., & Maczuga, S. (2017). Replicated evidence of racial and ethnic disparities in disability identification in US schools. *Educational Researcher, 46*(6), 305–322.

Nora, E. P., Lajoie, B. R., Schulz, E. G., Giorgetti, L., Okamoto, I., Servant, N., Piolot, T., Van Berkum, N. L., Meisig, J., & Sedat, J. (2012). Spatial partitioning of the regulatory landscape of the X-inactivation centre. *Nature, 485*(7398), 381–385.

Nuebler, J., Fudenberg, G., Imakaev, M., Abdennur, N., & Mirny, L. (2018). Chromatin organization by an interplay of loop extrusion and compartmental segregation. *Biophysical Journal, 114*(3), 30a.

Olivares-Chauvet, P., Mukamel, Z., Lifshitz, A., Schwartzman, O., Elkayam, N. O., Lubling, Y., Deikus, G., Sebra, R. P., & Tanay, A. (2016). Capturing pairwise and multi-way chromosomal conformations using chromosomal walks. *Nature, 540*(7632), 296–300.

Parada, L. A., McQueen, P. G., & Misteli, T. (2004). Tissue-specific spatial organization of genomes. *Genome Biology, 5*(7), 1–9.

Pascual, S., Bonafonte, A., & Serra, J. (2017). *SEGAN: Speech Enhancement Generative Adversarial Network*. Cornell University. arXiv preprint arXiv:1703.09452.

Pederson, T. (2011). The nucleolus. *Cold Spring Harbor Perspectives in Biology, 3*(3), a000638.

Peric-Hupkes, D., Meuleman, W., Pagie, L., Bruggeman, S. W., Solovei, I., Brugman, W., Gräf, S., Flicek, P., Kerkhoven, R. M., & van Lohuizen, M. (2010). Molecular maps of the reorganization of genome-nuclear lamina interactions during differentiation. *Molecular Cell, 38*(4), 603–613.

Quinodoz, S. A., Ollikainen, N., Tabak, B., Palla, A., Schmidt, J. M., Detmar, E., Lai, M. M., Shishkin, A. A., Bhat, P., & Takei, Y. (2018). Higher-order inter-chromosomal hubs shape 3D genome organization in the nucleus. *Cell, 174*(3), 744–757.e724.

Rao, S. S., Huang, S. C., St Hilaire, B. G., Engreitz, J. M., Perez, E. M., Kieffer-Kwon, K. R., Sanborn, A. L., Johnstone, S. E., Bascom, G. D., & Bochkov, I. D. (2017). Cohesin loss eliminates all loop domains. *Cell, 171*(2), 305–320.e324.

Rao, S. S., Huntley, M. H., Durand, N. C., Stamenova, E. K., Bochkov, I. D., Robinson, J. T., Sanborn, A. L., Machol, I., Omer, A. D., & Lander, E. S. (2014). A 3D map of the human genome at kilobase resolution reveals principles of chromatin looping. *Cell, 159*(7), 1665–1680.

Redolfi, J., Zhan, Y., Valdes-Quezada, C., Kryzhanovska, M., Guerreiro, I., Iesmantavicius, V., Pollex, T., Grand, R. S., Mulugeta, E., & Kind, J. (2019). DamC reveals principles of chromatin folding *in vivo* without crosslinking and ligation. *Nature Structural & Molecular Biology, 26*(6), 471–480.

Rivera-Mulia, J. C., Desprat, R., Trevilla-Garcia, C., Cornacchia, D., Schwerer, H., Sasaki, T., Sima, J., Fells, T., Studer, L., & Lemaitre, J. M. (2017). DNA replication timing alterations identify common markers between distinct progeroid diseases. *Proceedings of the National Academy of Sciences, 114*(51), E10972–E10980.

Rooijers, K., Markodimitraki, C. M., Rang, F. J., de Vries, S. S., Chialastri, A., de Luca, K. L., Mooijman, D., Dey, S. S., & Kind, J. (2019). Simultaneous quantification of protein–DNA contacts and transcriptomes in single cells. *Nature Biotechnology, 37*(7), 766–772.

Roukos, V., Voss, T. C., Schmidt, C. K., Lee, S., Wangsa, D., & Misteli, T. (2013). Spatial dynamics of chromosome translocations in living cells. *Science, 341*(6146), 660–664.

Sanders, M. E., Merenstein, D. J., Reid, G., Gibson, G. R., & Rastall, R. A. (2019). Probiotics and prebiotics in intestinal health and disease: From biology to the clinic. *Nature Reviews Gastroenterology & Hepatology, 16*(10), 605–616.

Senecal, A., Munsky, B., Proux, F., Ly, N., Braye, F. E., Zimmer, C., Mueller, F., & Darzacq, X. (2014). Transcription factors modulate c-Fos transcriptional bursts. *Cell Reports, 8*(1), 75–83.

Sima, J., Chakraborty, A., Dileep, V., Michalski, M., Klein, K. N., Holcomb, N. P., Turner, J. L., Paulsen, M. T., Rivera-Mulia, J. C., & Trevilla-Garcia, C. (2019). Identifying cis elements for spatiotemporal control of mammalian DNA replication. *Cell, 176*(4), 816–830.e8.

Strickfaden, H., Zunhammer, A., van Koningsbruggen, S., Köhler, D., & Cremer, T. (2010). 4D chromatin dynamics in cycling cells: Theodor Boveri's hypotheses revisited. *Nucleus, 1*(3), 284–297.

Sutherland, H., & Bickmore, W. A. (2009). Transcription factories: Gene expression in unions? *Nature Reviews Genetics, 10*(7), 457–466.

Van De Werken, H. J., Landan, G., Holwerda, S. J., Hoichman, M., Klous, P., Chachik, R., Splinter, E., Valdes-Quezada, C., Öz, Y., & Bouwman, B. A. (2012). Robust 4C-seq data analysis to screen for regulatory DNA interactions. *Nature Methods, 9*(10), 969–972.

Vernimmen, D., & Bickmore, W. A. (2015). The hierarchy of transcriptional activation: From enhancer to promoter. *Trends in Genetics, 31*(12), 696–708.

Wang, G., Wei, Y., Qiao, S., Lin, P., & Chen, Y. (2018). *Generalized Inverses: Theory and Computations* (Vol. *53*). Springer.

Wang, Q., Wang, Z., Awasthi, M. K., Jiang, Y., Li, R., Ren, X., Zhao, J., Shen, F., Wang, M., & Zhang, Z. (2016). Evaluation of medical stone amendment for the reduction of nitrogen loss and bioavailability of heavy metals during pig manure composting. *Bioresource Technology, 220*, 297–304.

Zhan, T., Rindtorff, N., & Boutros, M. (2017). WNT signaling in cancer. *Oncogene, 36*(11), 1461–1473.

Zhang, Y., & Dudko, O. K. (2016). First-passage processes in the genome. *Annual Review of Biophysics, 45*, 117–134.

Zheng, Q., & Lavis, L. D. (2017). Development of photostable fluorophores for molecular imaging. *Current Opinion in Chemical Biology, 39*, 32–38.

CHAPTER 7

Gene and Cell Division

AARADHYA BHATNAGAR,[1] SACHCHIT PATEL,[2] and ARPIT SHUKLA[2,*]

[1]*Indrashil University, Rajpur, Gujarat, India*

[2]*Indian Institute of Advanced Research, Koba Institutional Area, Gandhinagar, Gujarat, India*

*Corresponding author

ABSTRACT

Genetics in recent times has been richly developed through experimentation and research. Mendel was the first to set the outline for genetics even before chromosomes or genes were known. The distribution of genes is important for identifying coding and non-coding protein sequences. The Human Genome Project has set a framework in the field of genetics and is widely used in medicinal biology. Various methods for sequencing DNA are at the heart of diagnosing and treating diseases. In this chapter, we have also described cell division (CD), the final step of reproduction. CD is carried out in synchronized events, including different phases, which are discussed here. During meiosis, when crossing over occurs, both crossover and recombination frequencies are generated. These are used to determine gene mapping, which challenges Mendel's Laws of Inheritance. Techniques such as isotope labeling and flow cytometry are commonly used to study cell cycle phases. Cellular balance is necessary for continuance and is controlled by apoptosis and senescence. Genetics plays an important role in both processes. If these do not function accurately, aged or damaged cells may harm the host. In sum, there are promising new avenues in the study of genes and the cell cycle, with many more advances expected as technology evolves.

Chromosomal and Genomic Structures and Functions, Volume 1: Chromosomes and Cell Division.
Tariq Ahmad Bhat and Jameel M. Al-Khayri (Eds)

DOI: 10.1201/9781779640253-7

7.1 INTRODUCTION

The term "gene" is widely used these days. In the early 1860s, these inheritance traits were first noticed by Gregor Mendel. He used the term "cell elements" (now known as genes) to describe the phenomenon of inheritance using the pea plant (*Pisum sativum*). The proper definition of a gene is an evolving phenomenon due to advanced discoveries that provide different attributes, but for simplicity, according to the NCI Dictionary of Genetics Terms, it is defined as "the basic unit of heredity passed from parent to child." Each gene has a differing form known as an "allele." Diploid organisms, like humans, are called so due to the presence of two alleles. Each allele is derived from one of two parents. Alleles derived may be identical (known as the homozygous condition) or may differ (known as the heterozygous condition). In the case of unidentical or alternative forms, the allele that is able to express itself is known as the dominant allele, while the allele that is unable to express itself at the locus is known as the recessive allele. A chromosome is a filamentous structure composed of proteins and a solitary DNA molecule. It is located within the nucleus of the animal or plant cell. We shall learn the difference between genes and chromosomes in Table 7.1.

Each organism contains a specific set of genes that make it a functional living system. "Genome" is the entirety of an organism's genetic makeup. The particular set of genes in an individual is termed the "Genotype." The genotype of an organism determines the physical or morphological characteristics, which are referred to as "Phenotype." The human genome sequencing (Lander et al., 2001) provided a great understanding of the correlation between genotype and phenotype, which concerns human health (Gonzaga-Jauregui et al., 2012).

TABLE 7.1 Difference Between Gene and Chromosome

Gene	Chromosome
• Gene is found on chromosome.	• Chromosome are structures built with protein and DNA.
• Microscopically, genes are not visible.	• Chromosome can be observed under a microscope.
• Gene mutations are categorized in point mutations and frame shift mutations that is, insertion and deletion.	• Chromosome mutations lead to chromosomal disorders.

We shall talk about Mutation and their types in detail in upcoming topics.

7.2 HUMAN GENOME ARRANGEMENT

Understanding the arrangement of the human genome is a critical area of study in genetics and genomics, with implications for personalized medicine and advancements in the treatment of genetic disorders. The genome is organized into regions with different functions, including genes that code for proteins, non-coding regions that regulate gene expression, and repetitive sequences. The human genome is composed of DNA molecules, which are organized into 23 pairs of chromosomes. Each chromosome contains long sequences of genetic information, which are composed of four different nucleotide bases (adenine, guanine, cytosine, and thymine). These sequences of nucleotides make up genes, which code for the proteins that carry out most of the functions in our cells. We already know that mitochondria have an autonomous machinery; that is, they have their own genetic material. Therefore, the genetic information is enciphered into two genomes: nuclear and mitochondrial (Gonzaga-Jauregui et al., 2012). The human genome consists of genes that are not similar when it comes to size; the smallest being the testis determining factor, which is present on the Y chromosome and is essential for male character development, while the largest is dystrophin, present on the X chromosome. Gene size is a very important marker for the evolutionary aspect of human life; it is believed that genes with a longer length tend to have a high intronic load and are more evolutionarily conserved (Wolf et al., 2009). On the flip side, high expression, smaller proteins, and low intronic content are all correlated with shorter gene length (Urrutia and Hurst, 2003). However, with regard to recent advancements, we are yet to fully understand the intricacies and operation of the human genome (Simonti and Capra, 2015).

7.3 HUMAN GENE NUMBER

The number of genes generated by our genome is still not very well known. There have been many proposed methods to roughly estimate this number. However, these methods will not provide us with the accurate number of genes present (Table 7.2).

TABLE 7.2 Estimates of the Number of Genes in Human Genome Based on Various Methods

Method	Estimated Number of Genes
Experimental determination of protein-coding genes	19,000–20,000
Comparative genomics and gene prediction algorithms	19,000–25,000
Transcriptome sequencing and analysis	60,000–100,000
Analysis of alternative splicing and isoform diversity	90,000–200,000
Analysis of non-coding RNA transcripts	50,000–100,000

7.4 GENE DISTRIBUTION

The distribution of genes has been a subject of intense research for determining the location of protein-coding and non-coding sequences. There are some findings about gene distribution in the human genome.

1. **Protein-Coding Genes:** Protein-coding genes, which are extensively researched, constitute the most thoroughly studied category of genes in the human genome. These genes play a crucial role in generating a wide range of proteins that perform various activities within cells. The number of protein-coding genes in the human genome has been subject to varying estimations, but current research suggests that there are around 19,000–20,000 genes (International Human Genome Sequencing Consortium). The distribution of these genes is not uniform across the chromosomes, since certain chromosomes, such as chromosome 19, have a greater density of genes compared to others ("The human genome: organization and evolutionary history." *Annual Review of Genetics* vol. 29).

2. **Non-Coding RNA Genes:** Aside from protein-coding genes, the human genome also harbors many non-coding RNA genes that generate RNA molecules involved in regulating gene expression and other cellular activities. Identifying and characterizing non-coding RNA genes is more challenging compared to protein-coding genes. However, estimations indicate that there might be tens of thousands of distinct non-coding RNA genes in the human genome (Carninci et al., 2005).

3. **Repeats and Non-Coding Regions:** The majority of the human genome is composed of non-coding DNA, including repetitive sequences and other regions that do not code for genes. Gene expression and other cellular processes are regulated by these non-coding regions, but their function remains unclear. Repetitive sequences, in particular, are thought to be involved in chromosomal structure and stability, and mutations in repetitive regions have been linked to several genetic disorders (Smit AF, The origin of interspersed repeats in the human genome).

7.5 EXON CHARACTERISTIC

The majority of the human genome is composed of non-coding DNA, including repetitive sequences and other regions that do not code for genes. Gene expression and other cellular processes are regulated by these non-coding regions, but their function remains unclear. Repetitive sequences, in particular, are thought to be involved in chromosomal structure and stability, and mutations in repetitive regions have been linked to several genetic disorders (Smit AF, The origin of interspersed repeats in the human genome).

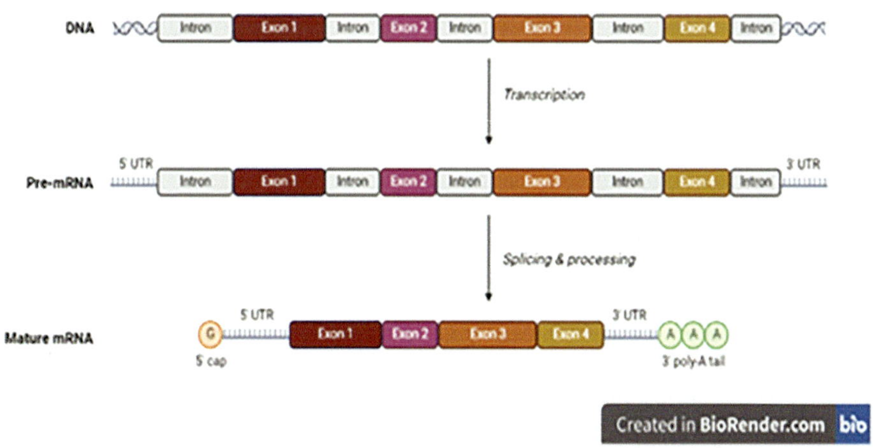

FIGURE 7.1 Schematic representation of exon and intron organization during gene expression.

7.6 HUMAN GENOME PROJECT

There is a dedicated project undertaken for sequencing the whole human genome, which is known as the Human Genome Project. This project mainly had three goals: (i) sequencing the genome as a whole; (ii) identification; and (iii) gathering optimized data analysis records for the same. The project started in 1990 and was completed by 2003.

7.6.1 METHODOLOGY

The HGP was a massive undertaking that involved hundreds of researchers from around the world. The project was divided into three main phases: mapping, sequencing, and analysis. The mapping phase involved identifying the locations of genes on the chromosomes, while the sequencing phase involved determining the order of the nucleotides (A, T, C, and G) in the DNA. The analysis phase involved interpreting the data and understanding the functions of the genes (Lander et al., 2001).

7.6.2 RESULTS

The HGP was a resounding success. The final sequence of the human genome was published in 2003, and it revealed a wealth of information about human biology and genetics. The HGP identified approximately 20,000–25,000 protein-coding genes in the human genome, as well as thousands of non-coding genes that play important regulatory roles (McPherson et al., 2001). The project also identified many disease-causing genes and provided new insights into the genetic basis of many complex diseases, including cancer, diabetes, and heart disease.

7.6.3 IMPACT

The HGP has had a profound impact on medicine and biology. It has led to the establishment of new tests for analysis and treatments for genetic diseases, as well as new approaches to personalized medicine. The HGP has also revolutionized the field of genetics, providing researchers with new tools and techniques for studying the genome and understanding its functions. The HGP has spawned many follow-up projects, including the 1000 Genomes Project, which aims to sequence the genomes of thousands of individuals from different populations, and the ENCODE project, which seeks to find all of the human genome's functional elements.

7.6.4 CONCLUSION

The Human Genome Project was a landmark achievement in the field of genetics and has had a profound impact on medicine and biology (Green et al., 2011). The project has given researchers a lot of information about the human genome and has opened up new possibilities and advances for the future (Figure 7.2).

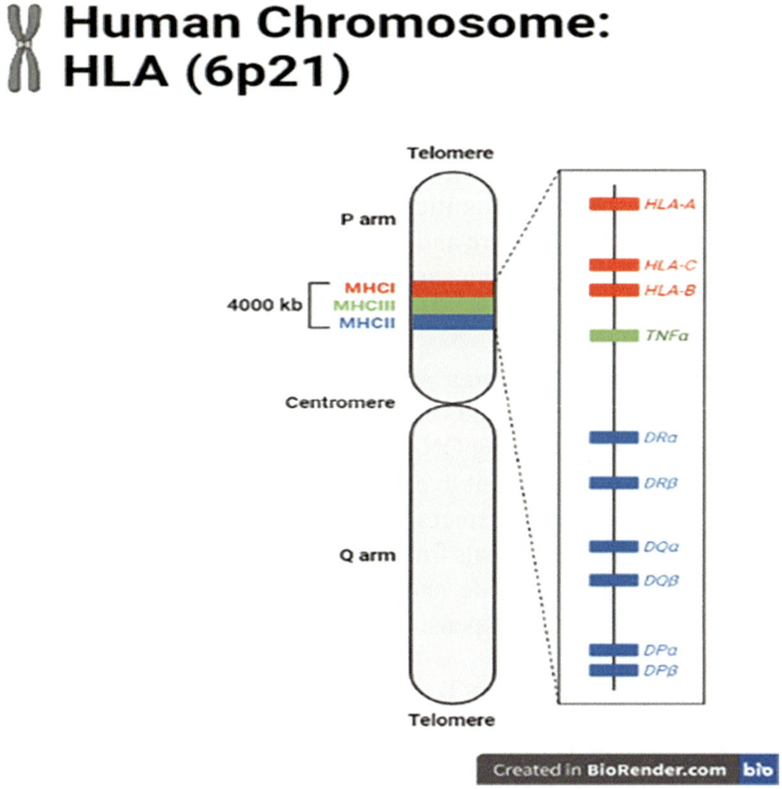

FIGURE 7.2 Schematic representation of the human leukocyte antigen (HLA) complex located on chromosome 6 at position 6p21.

7.7 METHODS FOR SEQUENCING OF DEOXYRIBONUCLEIC ACID (DNA)

Deoxyribonucleic Acid (DNA) sequencing is a method used to determine the exact order of nucleotides within a DNA molecule. There are some prevalent methods for DNA sequencing:

1. **Sanger Sequencing:** The Sanger method is a chain-termination technique that employs dideoxy nucleotides to halt the activity of DNA polymerase in extending the developing DNA chain. By employing a combination of regular and dideoxynucleosides, nucleotide sequences of varying lengths are generated. The fragments are subsequently segregated based on their size by gel electrophoresis, allowing the sequence to be determined by analyzing the band pattern. Sanger sequencing remains employed for the sequencing of limited DNA segments and for validating next-generation sequencing (NGS) outcomes (Sanger et al., 1977).

2. **Next-Generation Sequencing (NGS):** NGS technologies enable very efficient sequencing of several DNA fragments simultaneously, resulting in significantly accelerated sequencing rates and lower costs (Shendure and Ji, 2008). Multiple NGS systems exist, each possessing distinct advantages and disadvantages. Several of the most often utilized platforms include:

    i. **Illumina:** Illumina sequencing is based on the reversible terminator method, in which nucleotides are added one at a time to a growing DNA strand, and each nucleotide is labeled with a fluorescent dye. The fluorescent signal is detected and recorded by a camera, and the sequence is reconstructed by aligning the signals from each nucleotide. Illumina sequencing is highly accurate and can generate large amounts of data quickly and inexpensively (Goodwin, 2016).

    ii. **Ion Torrent:** Ion Torrent sequencing is based on the observation of hydrogen ions that are released upon the addition of a nucleotide to the growing DNA strand. The release of ions is detected by a semiconductor chip, and the sequence is determined by analyzing the pattern of ion signals. Ion

Torrent sequencing is fast and simple but has lower accuracy than Illumina sequencing.

iii. **PacBio:** PacBio sequencing employs a single-molecule real-time (SMRT) sequencing technique, enabling the generation of lengthy reads spanning tens of kilobases. This approach relies on the identification of light pulses produced by the integration of nucleotides into an expanding DNA strand. PacBio sequencing exhibits exceptional precision and is particularly advantageous for sequencing intricate segments of the genome. However, it is somewhat costlier than alternative NGS systems.

7.8 MUTATION

Mutation refers to a modification in the DNA sequence of any living creature. Mutations can arise due to mistakes in DNA replication, exposure to mutagens, or exposure to carcinogens, among other factors. The mutation that takes place in the somatic cell is not heritable; however, the mutation occurring in the germ cell can be transmitted to the progeny (Figure 7.3).

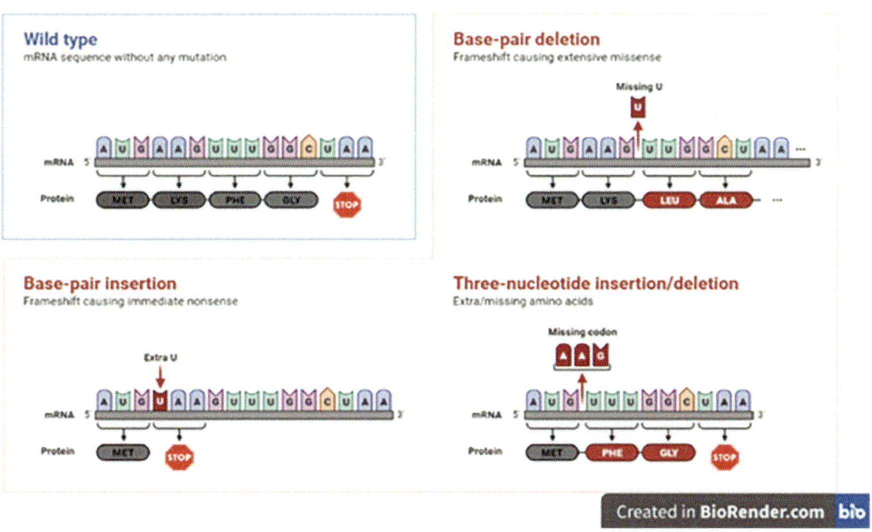

FIGURE 7.3 Types of genetic mutations and their effects on mRNA and protein synthesis.

Here are some of the most common types of mutations:

1. **Point Mutations:** Point mutations refer to alterations in a single nucleotide within the DNA sequence. They may be categorized into three distinct types:

    i. **Silent Mutations:** These mutations do not alter the sequence of amino acids in the protein encoded by the gene, as the mutated codon still corresponds to the identical amino acid as the initial codon.

    ii. **Missense Mutations:** Due to these mutations, a single amino acid in the protein sequence is altered since the mutated codon now corresponds to a different amino acid compared to the original codon

    iii. **Nonsense Mutations:** These mutations cause the protein sequence to have a premature stop codon, leading to the production of a shortened and non-functional protein (Mitelman, Johansson and Mertens, 2017).

2. **Insertions and Deletions:** These alterations entail the insertion or deletion of one or more nucleotides from the DNA sequence. These disruptions can cause a shift in the reading frame of the gene, resulting in a frameshift mutation that changes the sequence of amino acids in the protein (Pollard et al., n.d.).

3. **Chromosomal Rearrangements:** Chromosomal rearrangement pertains to alterations in the structure of a chromosome, encompassing deletions, duplications, inversions, or translocations. These modifications can occur as a result of mistakes in cellular replication, exposure to radiation or certain substances, or even spontaneously.

    i. **Deletions:** Sections of the chromosome are absent.

    ii. **Duplications:** It occur when segments of the chromosome are replicated, leading to the presence of more genetic material.

iii. **Inversions:** Chromosomal genetic material undergoes reversal.

iv. **Translocations:** It occur when segments of chromosomes undergo breakage and then get attached to a different chromosome (Alberts et al., 2015).

7.9 EPIGENETIC CHANGES

Epigenetics is the study of how the immediate environment and personal activities can influence gene function. Epigenetic inheritance plays a major role in the adaptation of an organism to today's ever-changing environment without altering the DNA sequence (Peschansky and Wahlestedt, 2014). Non-coding RNAs act as both direct and indirect modulators of epigenetic regulation (Peschansky and Wahlestedt, 2014).

Here are some examples of epigenetic changes and relevant references.

7.9.1 DEOXYRIBONUCLEIC ACID (DNA) METHYLATION

Methylation is the process of adding a methyl group to the cytosine base of a CpG dinucleotide, resulting in a chemical modification of DNA. DNA methylation can result in the suppression of gene expression and has been associated with several biological phenomena, such as embryonic development, X chromosome inactivation, and genomic imprinting (Jones and Liang, 2009).

7.9.2 HISTONE MODIFICATION

Histones are the proteins responsible for the packaging of DNA into a condensed form known as chromatin. Modifying histones has the capacity to impact the structure of chromatin and control the expression of genes. Histone acetylation is commonly associated with gene activation, whereas histone methylation can have either an activating or repressing effect on gene expression, depending on the precise residue being changed on the histone (Jenuwein and Allis, 2001).

7.9.3 NON-CODING RIBONUCLEIC ACID (RNA)

Non-coding RNA molecules, such as microRNAs and long non-coding RNAs, exert control over gene expression by attaching to messenger RNA (mRNA) and either suppressing protein synthesis or facilitating its breakdown. Non-coding RNAs have been linked to several biological processes, such as embryonic development, cancer, and neurodegenerative illnesses (Esteller, 2011).

7.9.4 CHROMATIN REMODELING

Chromatin remodeling is the modification of the chromatin structure to increase or decrease its accessibility to transcriptional machinery. This process may entail the activity of large protein complexes that either slide or eliminate nucleosomes, or the activity of ATP-dependent chromatin remodeling enzymes that utilize energy to relocate nucleosomes. Chromatin remodeling plays a crucial role in controlling gene expression and is involved in several biological processes, including differentiation and development (Ho and Crabtree, 2010).

7.9.5 ENVIRONMENTAL FACTORS

Epigenetic modifications can also be triggered by external factors such as food, stress, and exposure to pollutants. For example, the food of the mother during pregnancy can influence the DNA methylation patterns of the growing fetus. Additionally, exposure to pollutants such as cigarette smoke can lead to alterations in DNA methylation and histone modification in adult cells (Skinner, 2015).

7.10 CELL DIVISION (CD)

A cell goes through a specific sequence of events called the cell cycle before it can divide and produce copies of itself. The cell cycle is a tightly controlled process that guarantees proper cell division while keeping the cell's DNA intact.

There are the cell cycle's two primary stages, and they are known as interphase and mitosis. When a cell is not dividing, it is in a phase called interphase, during which it can expand and duplicate its DNA. The process by which DNA is duplicated and then split between two new cells is called mitosis.

There are three distinct stages: G1, S, and G2 during the interphase. A cell's growth and regular metabolic activities are carried out during the G1 phase. It is during the S phase that DNA replication occurs in the cell. The cell prepares for division during the G2 phase, when it synthesizes the proteins and other components it will need for the process.

A mitosis progresses through four phases called prophase, metaphase, anaphase, and telophase. Condensation of chromatin into visible chromosomes and disintegration of the nuclear sheath are hallmarks of prophase. During metaphase, chromosomes align along the cell's midline. When a cell enters anaphase, its sister chromatids become detached and are dragged to opposing ends. Telophase is the cell cycle's last stage, and it is when the nuclear envelope is reformed around the two sets of chromosomes and cytokinesis occurs.

Several checkpoints exist throughout the cell cycle to guarantee that the cell is properly prepared for the following phase. Cancer can originate from cells dividing incorrectly, which may occur if the cell cycle is not correctly controlled.

7.10.1 PHASES OF CELL CYCLE

Following the completion of the M phase, which encompasses the division of the cytoplasm and mitosis, the segregation of chromosomes and other cellular constituents occurs, resulting in the birth of each individual cell. In the S phase, the chromosomal DNA undergoes replication (synthesis phase). There are two more phases that occur between mitosis and the S phase. The period of time that elapses between mitosis and the initiation of DNA replication is referred to as the G1 phase, also known as the first gap phase. The period of time between the cessation of DNA replication and the initiation of mitosis is referred to as the G2 phase, sometimes known as the second gap phase. Each cycling cell undergoes both an M phase and an S phase. During the development of specific early embryos, the G1 and G2 phases are quite short and exhibit variability in their duration (Lodish et al., 2000).

7.10.2 INTERPHASE

Interphase, a stage in the cell cycle, is characterized by the cell preparing for division. The cell cycle's longest phase is the one that may be separated into three stages: G1, S, and G2. Interphase is a stage in the cell cycle where the cell undergoes growth, DNA replication, and prepares for either mitosis or meiosis (Becker et al., 2017).

1. **G1 Phase:** In the G1 phase, the cell undergoes growth and synthesizes new organelles and proteins. This phase also encompasses a sequence of regulatory processes that guarantee the cell's readiness to progress to the subsequent phase of the cell cycle. The G1 phase of interphase is highly varied and can last for a prolonged period of time, ranging from many days to a few hours, depending on the specific kind of cell (Cooper and Hausman, 2019).

2. **S Phase:** DNA replication takes place during the S phase. This implies that the genetic material of the cell is replicated in order to guarantee that each offspring cell has a complete set of chromosomes. In order to ensure that each chromosome is replicated just once, this process is rigorously regulated (National Center for Biotechnology Information, 2021).

3. **G2 Phase:** During the G2 phase, the cell experiences a concluding stage of growth and readies itself for cell division. The cell examines its DNA for any abnormalities and verifies that all necessary preparations for mitosis or meiosis are complete. The cell furthermore produces proteins that are essential for cellular division (Alberts et al., 2015).

Interphase is a crucial stage of the cell cycle, as it guarantees that the cell is adequately primed for CD. It is also a time when cells can respond to signals from their environment and adjust their behavior accordingly. For example, cells in the G1 phase can enter a state of quiescence (G0 phase) if they receive signals indicating that they should not divide (Compton, 2017).

7.10.3 M PHASE OR MITOSIS

M phase, or mitosis, is a kind of cellular division that produces two daughter cells with identical genetic composition. It is crucial for the development of new tissues, tissue growth, and tissue healing of organisms during development. During mitosis, the division of a parent cell occurs into two identical daughter cells through a series of stages that include prophase, metaphase, anaphase, and telophase. Specific cellular characteristics define each stage, such as the condensation of chromosomes, the chromosomes' placement at the metaphase plate, and during anaphase, the division of sister chromatids (Hanahan and Weinberg, 2011).

The process of mitosis is tightly regulated by a variety of signaling pathways, including the checkpoint kinase pathway, which ensures that cells do not enter anaphase until all chromosomes have the mitotic spindle correctly connected (BioRender.com).

Aside from its function in regular cellular division, mitosis can also have a crucial impact on the development of cancer. Dysfunctions in the process of cell division can cause instability in the structure of chromosomes and the buildup of genetic alterations, ultimately resulting in the formation of tumors (Alberts et al., 2002).

7.10.4 THESE ARE THE PHASES OF MITOSIS

7.10.4.1 INTERPHASE

During this phase, the cell is actively performing its functions, such as growth and DNA replication, without dividing. The DNA is present in the cell as the loose, uncondensed structure known as chromatin.

7.10.4.2 PROPHASE

The first mitotic stage during which chromatin condenses into chromosomes. Under a microscope, the chromosomes can be seen as distinct, compact structures. The nuclear envelope disintegrates, and the nucleolus, which produces ribosomes, vanishes. The centrosomes move in opposite directions toward the cell's poles, where they organize the microtubules (MTs) that separate the chromosomes.

7.10.4.3 PROMETAPHASE

During this phase, the spindle fiber-forming MTs affix to the kinetochores, which are chromosomal structures that aid during cell division in chromosomal alignment. The metaphase plate, the region where the chromosomes are located in the cell, starts to move toward the center.

7.10.4.4 METAPHASE

The chromosomes along the metaphase plate are in alignment, which is a fictitious line in the cell's center at this time. In order to align the chromosomes in readiness for separation, the spindle fibers attach to them at the kinetochores.

7.10.4.5 ANAPHASE

During this phase, the sister chromatids, which are identical copies of the chromosomes, are pulled apart in opposite directions by the spindle fibers. The spindle fibers that are not connected to the chromosomes slide past one another, further separating the poles, and the cell enlarges.

7.10.4.6 TELOPHASE

Currently, the chromosomes have migrated to the opposing poles of the cell. Following the disintegration of the spindle threads, a fresh nuclear membrane forms around each group of chromosomes, resulting in the division of two separate nuclei. The chromosomes undergo decondensation and revert to their chromatin state.

7.10.4.7 CYTOKINESIS

Cytokinesis is the cellular process of partitioning the cytoplasm and organelles between two daughter cells, resulting in their division. Animal cells undergo cytokinesis by the formation and progressive deepening of a cleavage furrow, resulting in the splitting of the cell into two. The centrosomes of plant cells evolve into cell plates, which then expand outward to facilitate the division of the cell into two equal halves.

Mitosis is a highly regulated process that ensures the accurate transmission of genetic information from one cell generation to the next (Lodish et al., 2000) (Figure 7.4).

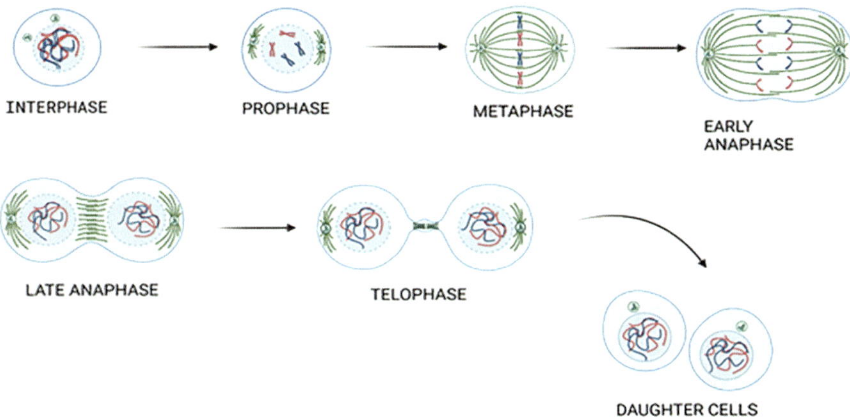

FIGURE 7.4 Stages of mitosis leading to cytokinesis.

7.10.5 MEIOSIS

Meiosis is a specific form of cellular division that occurs in animals that reproduce sexually. Meiosis, in contrast to mitosis, yields four daughter cells that are genetically distinct from one another. Meiosis I and Meiosis II are utilized to segregate sister chromatids and homologous chromosomes, respectively, in order to accomplish this. One pair of chromosomes from each parent begins meiosis when a diploid cell with two nuclei is divided into two. These homologous chromosomes link together and cross over during meiosis I, exchanging DNA segments in the process. This creates new combinations of genetic information that are different from either parent. Subsequently, the homologous chromosomes undergo separation and move towards opposite poles of the cell, resulting in the formation of two haploid daughter cells. Every descendant cell possesses a single pair of chromosomes, each consisting of two chromatids (Alberts et al., 2015).

In the process of meiosis II, the sister chromatids undergo separation and move to opposing poles of the cell, resulting in the formation of four haploid daughter cells. The daughter cells exhibit genetic diversity as a result of the stochastic arrangement of chromosomes in meiosis I and the segregation of sister chromatids in meiosis II. Meiosis is essential for sexual reproduction since it enables the formation of offspring with genetic diversity, which is desirable. It is also important for maintaining the correct number of chromosomes in a species, as the diploid number is restored when two haploid gametes fuse during fertilization (Campbell et al., 2017).

7.10.5.1 THE PROCESS OF MEIOSIS

Meiosis, a kind of cell division, occurs in organisms that reproduce sexually. The method necessitates two rounds of cell division to generate four haploid cells, each with half the number of chromosomes as the parent cell. Meiosis is an essential mechanism in sexual reproduction since it guarantees the presence of genetic variation among offspring. Below is an elaborate explanation of the process of meiosis:

1. **Prophase I:** It is the first stage of meiosis I. During this stage, homologous chromosomes pair up and undergo crossing over, a process in which they exchange genetic material. This genetic exchange facilitates the augmentation of genetic variety among progeny. The duration of Prophase I, the initial phase of meiosis I, is considerably longer and may require several days to complete (Alberts et al., 2015; BioRender.com).

2. **Metaphase I:** In metaphase I, the homologous pairs of chromosomes arrange themselves in a line down the center of the cell. The alignment is arbitrary, hence enhancing genetic variety (Campbell et al., 2017; Samanthi, 2019).

3. **Anaphase I:** In anaphase I, the homologous pairs of chromosomes undergo separation and are pulled toward opposite ends of the cell (Alberts et al., 2015).

4. **Telophase I:** It represents the last stage of meiosis I. At this stage, the nuclear membrane reassembles around each group of chromosomes, and the chromosomes migrate towards the opposite ends of the cell. Subsequently, the cell undergoes division, resulting in

the formation of two daughter cells, each with half the number of chromosomes compared to the parent cell (Campbell et al., 2017).

5. **Prophase II:** The second stage of meiosis, known as meiosis II, commences with prophase II, during which the chromosomes undergo condensation once again, and the nuclear membrane disintegrates (Alberts et al., 2015).

6. **Metaphase II:** In metaphase II, the chromosomes arrange themselves in a line in the center of the cell, much like in metaphase I. However, in contrast to metaphase I, the chromosomes are not arranged in homologous pairs (Campbell et al., 2017).

7. **Anaphase II:** In anaphase II, the sister chromatids of each chromosome undergo separation and are subsequently drawn towards opposing poles of the cell (Alberts et al., 2015).

8. **Telophase II:** It is the concluding phase of meiosis II. In this phase, the chromosomes migrate to opposing ends of the cell, and the nuclear membrane reassembles around each group of chromosomes. Subsequently, the cell undergoes division, leading to the formation of four haploid offspring cells, each possessing a distinct assortment of genetic material (Campbell et al., 2017).

Overall, the process of meiosis is a complex and important process that ensures genetic diversity among offspring. By following the steps outlined above, organisms are able to produce gametes that are genetically distinct from one another, increasing the chances of survival and adaptation (Lander and Botstein, 1987) (Figure 7.5).

7.11 CROSSOVER FREQUENCY AND RECOMBINATION FREQUENCY

7.11.1 CROSSOVER FREQUENCY

Genetic recombination takes place during meiosis. This process results in alterations to genetic traits during the production of gametes. Hence, genetic diversity arises among species. The occurrence of double-stranded

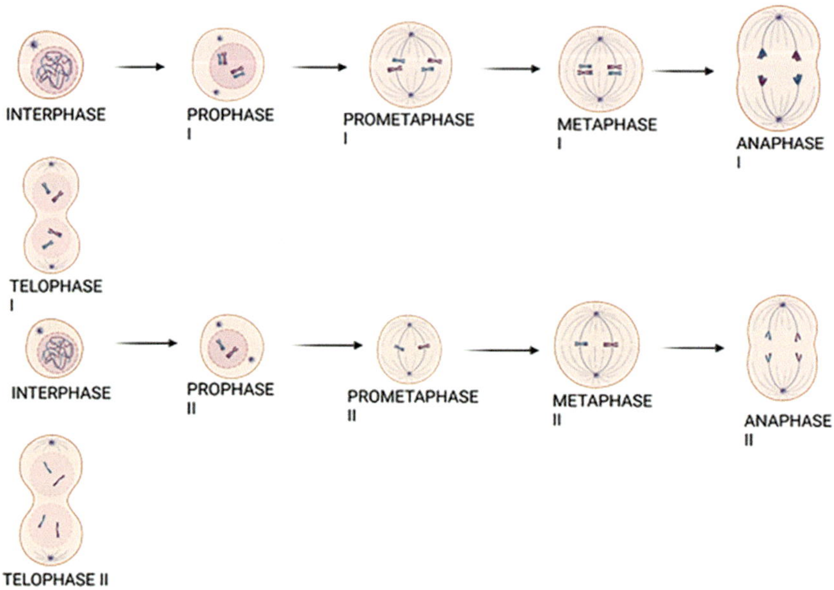

FIGURE 7.5 Sequential stages of meiosis I and meiosis II.

breaks signifies the initiation of the crossover process. This procedure is carried out by DNA topoisomerases. Double-stranded breaks lead to the creation of crossover genetic material. The frequency of crossover is governed by the rate at which the crossover takes place. The frequency of genetic recombination is contingent upon the genetic distance between the genes. The genetic map of a gene is established based on the frequency of crossover events. Crossover frequency may be categorized as homozygous crossover and heterozygous crossover (International Human Genome Sequencing Consortium, 2001).

7.11.2 RECOMBINATION FREQUENCY

Recombination is the process by which genes cross over in a heterozygous manner. The recombination frequency is the frequency or rate at which recombination occurs during crossover. The increase in recombination frequency suggests a pattern of offspring variation. The rate of recombination or heterozygous crossover is determined by the distance between the genes. As a result, the frequency of recombination increases as the distance between the genes decreases. This increases the genetic linkage between the genes (International Human Genome Sequencing Consortium, 2001).

What resemblances and contrasts exist between crossover and recombination frequencies?

- Both occur during meiosis.
- Crossing over occurs when generating crossover and recombination frequencies.
- The frequency of crossover and recombination is determined by the distance between the two genes.
- Both types of frequencies are used to determine genetic linkage and gene mapping (International Human Genome Sequencing Consortium, 2001).

7.11.3 GENE MAPPING

Gene mapping is the process of determining the location and relative distances between genes on a chromosome. There are several methods for gene mapping, including linkage analysis, physical mapping, and genome-wide association studies (GWAS). Here are some references on gene mapping:

7.11.3.1 LINKAGE ANALYSIS

The genetic study known as linkage determines the precise chromosomal position of genes that are accountable for a certain characteristic. This approach entails examining the co-segregation of genetic markers with the characteristic of interest within a family. Linkage analysis is a method used to locate genes that are responsible for hereditary illnesses (Visscher et al., 2012).

7.11.4 PHYSICAL MAPPING

Physical mapping is a method that identifies the physical location of genes on a chromosome by analyzing the DNA sequence. This method involves creating the genome's physical map, which can be used to identify the location of specific genes. Physical mapping is commonly used to study complex genetic disorders (Morgan Unit(s), 2006).

7.11.5 *GENOME-WIDE ASSOCIATION STUDIES (GWAS)*

Genome-wide association studies (GWAS) are a technique used to detect genetic variations linked to a certain characteristic or illness. GWAS involve comparing the DNA of individuals with and without the trait of interest to identify genetic variations that are more common in individuals with the trait. GWAS have been used to identify genes associated with many common disorders like heart disease and diabetes (Morgan, 2007).

7.11.6 *MORGAN'S UNIT*

Morgan unit(s) is a unit of genetic recombination measured by crossing over (on average, one crossing over per meiosis). It specifies the separation between two genes. One centimorgan (cM) equals 1% recombination (Shapiro, 2003).

7.12 TECHNIQUES USED TO STUDY CELL CYCLE

The complex process of the cell cycle is essential for the proliferation and maturation of all living creatures. There are several techniques used to study the cell cycle, including microscopy, flow cytometry, biochemical assays, and genetic analysis. Here are some of the most commonly used techniques:

7.12.1 *MICROSCOPY*

Microscopy is an effective instrument for investigating the cell cycle. Researchers may utilize it to visually represent the many stages of the cell cycle and the structures implicated in CD. Techniques such as fluorescence microscopy and confocal microscopy can be used to label and track specific molecules and structures within cells during the cell cycle (Alberts et al., 2002).

7.12.2 *FLOW CYTOMETRY*

Flow cytometry is a high-throughput and precise method used by researchers to evaluate a huge quantity of cells efficiently. This technique involves passing cells through a flow cytometer, which measures various characteristics of the cells, such as their size, shape, and DNA content. Flow cytometry can be used to identify and analyze cells in different stages of the cell cycle (Sherr and Roberts, 1999).

7.12.3 *BIOCHEMICAL ASSAYS*

Biochemical assays are used to measure the activity of specific proteins and enzymes that are a part of the cell cycle. For example, cyclin-dependent kinase (CDK) assays are employed to quantify the activity of CDKs, which play a pivotal role in regulating the cell cycle (Galluzzi et al., 2018).

7.12.4 *GENETIC ANALYSIS*

Genetic analysis can be used to study the cell cycle by manipulating the genes involved in CD. For example, researchers can use gene knockout or RNA interference techniques to silence the expression of specific genes involved in the cell cycle and study the effects on CD (Kalkavan and Green, 2018).

7.13 APOPTOSIS AND SENESCENCE

Apoptosis, or programmed cell death, is a highly controlled process in which cells undergo a series of biochemical and morphological changes that ultimately lead to their own destruction. It plays a pivotal role in several biological processes, including embryonic development, tissue homeostasis, and immune system function. The disruption of apoptosis has been linked to several human illnesses, such as cancer, neurological disorders, and autoimmune diseases (Elmore, 2007).

The initiation of apoptosis is marked by the activation of a group of cysteine proteases known as caspases. These caspases cleave different cellular proteins, resulting in the distinct morphological alterations seen in apoptotic cells, including cell shrinkage, chromatin condensation, and fragmentation of the nucleus into apoptotic bodies. The regulation of the apoptotic process is intricately controlled by a sophisticated network of signaling channels, which may be activated by either extrinsic or intrinsic signals, depending on the kind of stimuli (Evan and Vousden, 2001).

7.14 IMPORTANCE OF APOPTOSIS

The importance of apoptosis has been extensively studied in different branches of biology and medicine. For example, in embryonic development, apoptosis is crucial for sculpting and refining the shape and size of developing organs and tissues. In the immune system, apoptosis is critical for the elimination of self-reactive lymphocytes and the maintenance of immune tolerance (van Doeselaar and Burgering, 2018). In cancer, dysregulation of apoptosis can lead to uncontrolled cell proliferation and resistance to chemotherapy or radiation therapy. Apoptosis is hypothesized to contribute to the death of neurons in brain areas affected by neurodegenerative illnesses, including Alzheimer's and Parkinson's disease.

7.14.1 SENESCENCE

Senescence is an inherent cellular reaction to chronic stress, characterized by a permanent cessation of the cell cycle, hence preventing the buildup of damaged cells. Various causes can initiate senescence, including telomere erosion, DNA damage, and oncogene stimulation. This mechanism triggers the activation of certain proteins, such as p53/p21 or p16/RB, which ultimately lead to the cessation of the cell cycle. Senescent cells are often identified by the presence of elevated levels of the cell cycle inhibitors p21 and/or p16.

Senescence can be triggered by several internal and external factors, including DNA damage, oxidative stress, telomere shortening, inflammation, and exposure to certain chemicals or poisons. The activation of senescence can lead to the propagation of this state to adjacent cells, a phenomenon referred to as the senescence-associated secretory phenotype (SASP).

The SASP is characterized by the release of several pro-inflammatory cytokines, chemokines, and growth factors, which can have both beneficial and detrimental effects on neighboring tissues and organs. On one hand, the SASP can help to eradicate damaged cells and encourage tissue restoration. On the other hand, excessive SASP activity can lead to chronic inflammation, tissue dysfunction, and an increased risk of age-related diseases.

Researchers are exploring various strategies to target senescence and SASP activity to promote healthier aging and extend lifespan. One approach is to create medication that removes senescent cells with precision, known as senolytics. Another approach is to target specific signaling pathways or molecules that are involved in senescence and SASP activity, such as the mTOR pathway, NAD^+ levels, or the sirtuin proteins.

Overall, senescence is a complex biological process that is crucial for the aging process and disorders associated with aging. By better understanding the mechanisms of senescence, researchers hope to develop new interventions that can slow down or even reverse some of the negative effects of aging.

7.15 MAXIMUM MUTATION PROBABILITY

The probability of mutation is generally highest during DNA replication, which happens during the cell cycle's S phase (synthesis phase). During DNA replication, errors in copying the genetic material can occur, leading to mutations. These mutations can be introduced as a result of errors made by the DNA polymerase enzymes responsible for replicating the DNA or from exposure to external mutagens such as chemicals or radiation.

It's important to note that not all mutations result in harmful effects. Some mutations can be neutral or even beneficial, while others can lead to diseases or other adverse effects. The frequency of mutations can be influenced by various factors, including the age of the organism, exposure to environmental factors, and the presence of certain genetic conditions.

7.16 FUTURE PERSPECTIVE

The study of genes and the cell cycle is an important and constantly evolving field of biology, and there are many exciting developments on the horizon. Here are a few potential areas of progress in the near future:

1. **Gene Editing:** The development of CRISPR-Cas9 technology has opened up exciting possibilities for targeted gene editing. This technique allows scientists to selectively modify or remove specific genes from an organism's DNA, which has a wide range of potential applications in industries like medicine, agriculture, and biotechnology.

2. **Personalized Medicine:** With advances in genetics, it may soon be possible to tailor medical treatments to an individual's genetic profile. This could include developing drugs that are specifically targeted to a patient's genetic makeup or using genetic testing to diagnose and treat diseases at an early stage.

3. **Cell Cycle Regulation:** The cell cycle is a complex process that controls the growth and division of cells, and understanding its regulation is key to developing treatments for diseases such as cancer. Researchers are working to identify the key molecular players involved in the cell cycle and to develop new drugs that can manipulate these processes.

4. **Stem Cell Research:** Due to their capacity to develop into a variety of cell types, stem cells hold great promise as a tool for regenerative medicine. Researchers are working to understand the molecular processes involved in stem cell differentiation and to develop new techniques for growing and manipulating stem cells.

Overall, the study of genes and the cell cycle is a rapidly advancing field with many exciting possibilities for the future. As technology and research methods continue to evolve, we can expect to see many new breakthroughs in our understanding of these fundamental biological processes.

KEYWORDS

- **apoptosis and senescence**
- **cell division**
- **CRISPR-Cas9 gene editing**
- **genetics**
- **human genome project**
- **personalized medicine**
- **stem cell research**

REFERENCES

Alberts, B., Johnson, A., Lewis, J., Raff, M., Roberts, K., & Walter, P. (2015). *Molecular Biology of the Cell* (6th ed.). Garland Science.

Alberts, B., Johnson, A., Lewis, J., Raff, M., Roberts, K., & Walter, P. (2002). *Molecular Biology of the Cell* (4th ed.). New York: Garland Science. Retrieved from: https://www.ncbi.nlm.nih.gov/books/NBK26823/ (accessed on 18 March 2025).

Alberts, B., Johnson, A., Lewis, J., Raff, M., Roberts, K., & Walter, P. (2015). *Molecular Biology of the Cell* (6th ed.). New York: Garland Science.

Alberts, B., Johnson, A., Lewis, J., Raff, M., Roberts, K., & Walter, P. (2002). *Molecular Biology of the Cell* (4th ed.). Garland Science.

Anzai, T., & Shiina, T. (2015). The molecular evolution of Major Histocompatibility Complex class I genes in primates. *Critical Reviews in Immunology, 35*(2), 119–134.

Becker, W. M., Kleinsmith, L. J., Hardin, J., & Bertoni, G. P. (2017). *The World of the Cell* (9th ed.). Pearson.

BioRender.com. Created with BioRender.com (https://www.biorender.com/).

Campbell, N. A., Reece, J. B., Urry, L. A., et al. (2017). *Biology* (11th ed.). Pearson.

Carninci, P., Kasukawa, T., Katayama, S., et al. (2005). The transcriptional landscape of the mammalian genome. *Science, 309*(5740), 1559–1563.

Compton, D. A. (2017). Mechanisms of chromosome alignment and segregation during mitosis. *Annual Review of Biochemistry, 86*, 85–108.

Cooper, G. M., & Hausman, R. E. (2019). *The Cell: A Molecular Approach* (8th ed.). Sinauer Associates.

Elmore, S. (2007). Apoptosis: A review of programmed cell death. *Toxicologic Pathology, 35*(4), 495–516. https://doi.org/10.1080/01926230701320337.

Esteller, M. (2011). Non-coding RNAs in human disease. *Nature Reviews Genetics, 12*(12), 861–874.

Evan, G. I., & Vousden, K. H. (2001). Proliferation, cell cycle and apoptosis in cancer. *Nature, 411*(6835), 342–348. https://doi.org/10.1038/35077213.

Galluzzi, L., Vitale, I., Aaronson, S. A., et al. (2018). Molecular mechanisms of cell death: Recommendations of the Nomenclature Committee on Cell Death 2018. *Cell Death & Differentiation, 25*(3), 486–541. https://doi.org/10.1038/s41418-017-0012-4.

GC, A. T. (2002). The human genome: Organization and evolution. *Annual Review of Genomics and Human Genetics, 3*, 297–320.

Gonzaga-Jauregui, C., Lupski, J. R., & Gibbs, R. A. (2012). Human genome sequencing in health and disease. *Annual Review of Medicine, 63*, 35–61.

Goodwin, S., McPherson, J. D., & McCombie, W. R. (2016). Coming of age: Ten years of next-generation sequencing technologies. *Nature Reviews Genetics, 17*(6), 333–351.

Green, E. D., & Guyer, M. S. (2011). Charting a course for genomic medicine from base pairs to bedside. *Nature, 470*(7333), 204–213.

Hanahan, D., & Weinberg, R. A. (2011). Hallmarks of cancer: The next generation. *Cell, 144*(5), 646–674.

Ho, L., & Crabtree, G. R. (2010). Chromatin remodeling during development. *Nature, 463*(7280), 474–484.

International Human Genome Sequencing Consortium. (2001). Initial sequencing and analysis of the human genome. *Nature, 409*(6822), 860–921.

Jenuwein, T., & Allis, C. D. (2001). Translating the histone code. *Science, 293*(5532), 1074–1080.

Jones, P. A., & Liang, G. (2009). Rethinking how DNA methylation patterns are maintained. *Nature Reviews Genetics, 10*(11), 805–811.

Kalkavan, H., & Green, D. R. (2018). MOMP, cell suicide as a BCL-2 family business. *Cell Death & Differentiation, 25*(1), 46–55. https://doi.org/10.1038/cdd.2017.188.

Katsanis, N. (2004). The continuum of causality in human genetic disorders. *Genome Biology, 5*(4), 229.

Lander, E. S., & Botstein, D. (1987). Mapping Mendelian factors underlying quantitative traits using RFLP linkage maps. *Genetics, 121*(1), 185–199.

Lander, E. S., Linton, L. M., Birren, B., et al. (2001). Initial sequencing and analysis of the human genome. *Nature, 409*(6822), 860–921.

Lodish, H., Berk, A., Zipursky, S. L., Matsudaira, P., Baltimore, D., & Darnell, J. (2000). *Molecular Cell Biology* (4th ed.). New York: W. H. Freeman. Retrieved from: https://www.ncbi.nlm.nih.gov/books/NBK21659/ (accessed on 18 March 2025).

Makałowski, W. (2001). The human genome structure and organization. *Acta Biochimica Polonica, 48*(3), 587–598.

McPherson, J. D., Marra, M., Hillier, L., et al. (2001). A physical map of the human genome. *Nature, 409*(6822), 934–941.

Mitelman, F., Johansson, B., & Mertens, F. (2017). The impact of translocations and gene fusions on cancer causation. *Nature Reviews Cancer, 17*(4), 233–245.

Morgan Unit(s). (2006). In *Encyclopedic Reference of Genomics and Proteomics in Molecular Medicine.* Springer, Berlin, Heidelberg. https://doi.org/10.1007/3-540-29623-9_7875 (accessed on 18 March 2025).

Morgan, D. O. (2007). *The Cell Cycle: Principles of Control.* New Science Press.

National Center for Biotechnology Information. (2021). *Bookshelf: The Cell: A Molecular Approach* (2nd ed.).

Pollard, T. D., Earnshaw, W. C., Lippincott-Schwartz, J., & Johnson, G. T. (n.d.). *Cell Biology* (3rd ed.).

Sakharkar, M. K., Chow, V. T. K., & Kangueane, P. (2004). Distributions of exons and introns in the human genome. *In Silico Biol., 4*(4), 387–393.

Samanthi. (2019). *Difference Between Crossover Frequency and Recombination Frequency.* DifferenceBetween.com. https://www.differencebetween.com/difference-between-crossover-frequency-and-recombinationfrequency/ (accessed on 18 March 2025).

Sanger, F., Nicklen, S., & Coulson, A. R. (1977). DNA sequencing with chain-terminating inhibitors. *Proceedings of the National Academy of Sciences, 74*(12), 5463–5467.

Shapiro, H. M. (2003). *Practical Flow Cytometry.* Wiley-Liss.

Shendure, J., & Ji, H. (2008). Next-generation DNA sequencing. *Nature Biotechnology, 26*(10), 1135–1145.

Sherr, C. J., & Roberts, J. M. (1999). CDK inhibitors: Positive and negative regulators of G1-phase progression. *Genes & Development, 13*(12), 1501–1512. https://doi.org/10.1101/gad.13.12.1501.

Simonti, C. N., & Capra, J. A. (2015). Evolution of lysine acetylation in the RNA polymerase II C-terminal domain. Published in BMC Evolutionary Biology in 2015.

Skinner, M. K. (2015). Environmental epigenetics and a unified theory of the molecular aspects of evolution: A neo-Lamarckian concept that facilitates neo-Darwinian evolution. *Genome Biology and Evolution, 7*(5), 1296–1302.

Smit, A. F. (1996). The origin of interspersed repeats in the human genome. *Current Opinion in Genetics & Development, 6*(6), 743–748.

Urrutia, A. O., & Hurst, L. D. (2003). The signature of selection mediated by expression on human genes. *Genome Research, 13*, 2260–2264.

van Doeselaar, S., & Burgering, B. M. T. (2018). FOXOs maintaining the equilibrium for better or for worse. *Current Topics in Developmental Biology, 127*, 49–103. https://doi.org/10.1016/BS.CTDB.2017.10.003.

Visscher, P. M., Brown, M. A., McCarthy, M. I., & Yang, J. (2012). Five years of GWAS discovery. *The American Journal of Human Genetics, 90*(1), 7–24.

Wolf, Y. I., Novichkov, P. S., Karev, G. P., Koonin, E. V., & Lipman, D. J. (2009). The universal distribution of evolutionary rates of genes and distinct characteristics of eukaryotic genes of different apparent ages. *Proceedings of the National Academy of Sciences, 106*, 7273–7280.

CHAPTER 8

Sex Determination

FIZA TALAT, SARA ZAFAR,* and UMBREEN BIBI

Government College University, Faisalabad, Pakistan

*Corresponding author

ABSTRACT

The process of "sex determination" involves the differentiation and separation of the structures that produce male and female gametes. Sex determination generally addresses the genetic and environmental factors that impact sex differentiation, whereas sex differentiation is the physical manifestation of these events with regard to ovarian or testicular development. Thus, the main focus of sex determination is on gonadal sex and the physiological mechanisms that underpin gonadal function and development. Different regulatory mechanisms and genes are used by sex determination systems to determine activities in males and females that govern a common gene switch that controls sexual development. Gametes that are ambiguous undergo cyto-differentiation to produce sexual dimorphism in male and female gametes. These genes are found in sex chromosomes.

8.1 INTRODUCTION

The ultimate purpose of sex determination is to enhance heterozygosity of genes and decrease inbreeding by encouraging outcrossing, thereby increasing genetic fitness. Despite the fact that most plants are hermaphrodites—that is, they produce both pistils and stamens on a single flower—they have developed a variety of strategies and processes to enable outcrossing (Chuck, 2010; Connallon et al., 2022; Mitiku and Tujuba,

Chromosomal and Genomic Structures and Functions, Volume 1: Chromosomes and Cell Division.
Tariq Ahmad Bhat and Jameel M. Al-Khayri (Eds)

DOI: 10.1201/9781779640253-8

2022; Sultanova et al., 2023; Xirocostas et al., 2020). These include a wide range of mechanisms, such as genetic self-incompatibility and heterostyly, a hermaphrodite flower characterized by varying lengths of stamens and styles that inhibit self-fertilization. Several model systems have been used to determine the numerous processes that regulate the determination of sex. For instance, homosporous ferns like *Ceratopteris richardii* have undergone extensive research on both the genetics of sex determination, and loci responsible for the development of male, female, and hermaphrodite sex types have been identified (Atallah et al., 2018). Other basal plants, such as bryophytes and lycophytes, also undergo sex determination while having radically different reproductive systems than angiosperms. Sex determination in flowering plants has been extensively researched in *Silene latifolia*, which contains an X and Y chromosomal system similar to that seen in many mammals (Marais et al., 2018). The wide range of sexual expression found in angiosperms contradicts the theory that the process of determining sex is driven by a single mechanism that has probably evolved several times independently (Wright et al., 2016). The cloning of the genes responsible for determining sex in monoecious and dioecious plants, and also the discovery that these genes function through a number of distinct pathways, have confirmed this.

There is a lot of heterogeneity in plants regarding the location, timing, and process of this choice. Perfect or hermaphrodite flowers, in which the male and female reproductive parts develop in close proximity to one another inside the same flower, are produced by the majority of angiosperm species (Paterno et al., 2020). The dioecious species, on the other hand, are at the opposite extreme of the spectrum, with each individual producing solely male or female reproductive organs. This variance in sex expression suggests that plants have a wide variety of sex-determining processes, each of which has developed in tandem with the ecological niche that each species occupies (Malabadi et al., 2023).

8.2 SEX DETERMINATION IN THE PRIMARY LAND PLANT LINEAGES

A significant turning point in the history of life was the development of sexual reproduction, which happened long before plants made their first landfall some 470 million years ago (Meissner, 2021). By then, the earliest forms of land plants had developed distinct sex organs (archegonia and antheridia) to generate female and male gametes (spermatozoids and eggs). The diploid and haploid generations (the sporophytes and gametophytes) of the first terrestrial plants had also previously evolved (Suzuki, 2009). The differentiation of sex organs into tissues that produce eggs or sperm takes place during the haploid gametophytic phase in all land plants. The phase that makes the decision and the manner in which it is implemented differ. For a brief overview of how sex is determined in terrestrial plants (Renner and Müller, 2022), three groupings can be used to conveniently categorize their diversity: a nonvascular group that includes mosses, liverworts, and hornworts (Renner et al., 2017); the vascular group (seedless plants), including monilophytes and lycophytes (Sessa and Der, 2016); and the group of vascular (seed) plants, such as angiosperms (flowering plants) and gymnosperms (Liu et al., 2022). Gametophytes in non-seed plants can be either female or male (mostly seen in "dioecious") (Leite et al., 2021) species, where individual members produce either archegonia or antheridia separately, or in "monoecious" species, they might be hermaphrodites (Zhou et al., 2022), with the same individuals exhibiting both archegonia and antheridia. Here, the differentiation of tissues and cells in several body parts of the identical individual determines sex. Female and male gametophytes separate from identical-sized spores in dioecious species (Fu et al., 2020). However, sexual dimorphism can go too far. Male dwarf gametophytes can occasionally grow on the larger female gametophytes. The (V and U) sex chromosomes dictate sex in gametophytes in many bryophytes (Cronk, 2022). The function and goal of sex determination in ferns and lycophytes vary between "homosporous" species, where sporophytes generate spores that are all the same size, similar to those of bryophytes, and "heterosporous" species that have both larger "megaspores" and smaller "microspores" produced by sporophytes. The gametophytes in certain homosporous ferns have both antheridia and archegonia, making them functionally hermaphroditic. In these situations, determining an individual's sex is a matter of their tissues and cells differentiating into distinct male and female "gametangia." In other homosporous species, gametophytes may become unisexual (generating only archegonia or only antheridia) (Marchant, 2019). Hence, the process responsible for determining sex here focuses on distinct, unique gametophytes. Unlike some of the mosses, gametophytes differentiate in response to environmental signals sensed shortly after the spores germinate, not in reaction to genetic variation. In fact, no ferns or lycophytes have had their sex chromosomes characterized (Marchant et al., 2022). The size of the spores that causes the gametophytes' germination determines the sex of "heterosporous" ferns and lycophytes (Marchant et al., 2019). The spore-producing "sporangia" are regulated by the sporophyte to generate either larger "megaspores," which form female (mega) gametophytes, or smaller "microspores," which form male (micro) gametophytes, while the gametophytes in this case are genetically identical. Heterospory has independently developed in the evolutionary lineage that produced seed-bearing plants as well as in ferns, vascular terrestrial plants, and lycophytes (Pannell, 2017). The gametophytic phase is significantly more limited and altered in angiosperms as well as in gymnosperms. Thus, the sporophytic structures in which the male and female structures are differentiated give rise to gametophytes (Cronk, 2022). In living seed plants, the gametophytes completely transfer the role of determining sex to their diploid sporophyte (hosts). Ultimately, seed plant sporophytes can be either male or female, producing only pollen or ovules, or they can be functionally hermaphroditic, producing both ovules (which have modified female gametophytes) and pollen (which are modified and significantly reduced male gametophytes). Unlike non-seed plants, the sporophyte is responsible for determining the sex of gametophytes in seed plants, which are consistently either female or male (Renner and Müller, 2022). Therefore, in seed plants, the developmental mechanism of sex determination determines the types of gametophytes that a particular diploid sporophyte individual produces, as well as their distribution and proportions on the plant (i.e., the sex allocation) (Tanurdzic and Banks, 2004) (Table 8.1).

TABLE 8.1 Different Sex Expression Types in Various Pant Species N Sex Determination

Sex Expression Types in Plants	Phenotype	Description	Examples	References
Monoecious	♀ ♂	Separate flowers but on same plant	*Cucumis sativus*	Kiełkowska and Dziurka (2021)
Gynomonoecious	♀ ⚥	Hermaphrodite and female flower on same plant	*Acer campestre*	Jabbour et al. (2022)
Andromonoecious	♂ ⚥	Hermaphrodite and male flower on same plant	*Cucumis melo*	Rodriguez-Granados et al. (2017)

TABLE 8.1 *(Continued)*

Sex Expression Types in Plants	Phenotype	Description	Examples	References
Polygamous (Trimonoecious)	♀ ♂ ⚥	Hermaphrodite, male and female flower on same plant	*Musa acuminata*	Martin et al. (2017)
Dioecious	♂ or ♀	Separate plants that is, either male plant or female plant	*Phoenix dactylifera*	Intha and Chaiprasart (2018)
Gynodioecious	♀ or ⚥	Either hermaphrodite and female flower on same plant	*Silene italica*	Lafuma and Maurice (2006)
Androdioecious	♂ or ⚥	Either hermaphrodite and female flower on same plant	*Castilla elastica*	Wolf and Takebayashi (2004)
Subdioecious (Trioecious)	♀ and ♂ and ⚥	Hermaphrodite, male or female flower on same plant	–	Leite et al. (2021)
Hermaphorodite	⚥	Perfect flowers i.e., have both female and male organ	*Carica papaya*	Ávila-Hernández et al. (2023)

8.2.1 THE BRYOPHYTES

The modern hornworts, mosses, and liverworts are among the group of plants known as bryophytes. Their sporophyte generation (diploid and spore-producing) is smaller, short-lived, inconspicuous, and parasitic on the gametophyte (independent and gamete-producing) generation. In this group, the independent phase of the plant life cycle is the haploid gametophyte (Patiño and Vanderpoorten, 2018). However, all bryophytes produce one kind of spore, which is homosporous, and are characterized by species that are either homothallic (i.e., individuals have both female and male gametangia or sex organs) or heterothallic (i.e., having either female or male gametangia or sex organs) (Leslie and Bonacorsi, 2022). For example, certain species of the genus *Micromitrium* exhibit extreme sexual dimorphism, in which the dwarf male gametophyte grows on the leaves of the larger female gametophyte. Heterothallism, or unisexuality, has been linked to the presence of sex chromosomes in numerous bryophyte species. They are the only identified homosporous plant group that determines sex using sex chromosomes, despite the fact that the degree of sex chromosomes and unisexuality in the bryophytes has not been evaluated systematically (Charlesworth, 2022). Among bryophyte species, the heterothallic liverwort such as *Marchantia polymorpha* has been the subject of most research on bryophyte sex determination nowadays. The thalli, or vegetative gametophytes, of this species are similar in appearance, although the morphology of the sexual structures allows females and males to be easily identified from one another. A gametophyte can be an archegoniophore in females (when sex organs or gametangia are produced on the thalloid upper surface) or an antheridiophore in males (when the gametophyte has stalked branches on which gametangia are produced) (Mitiku and Tujuba, 2022). Archegoniophores produce archegonia, which form eggs, while antheridiophores produce antheridia, which form sperm. Males have no X chromosome while having a single Y chromosome; on the other hand, females have no Y chromosome while having a single X chromosome. These cytologically different sex chromosomes determine the sex of every haploid gametophyte. In bryophytes, *Marchantia* is considered a model organism suitable for genomics-style examinations due to its relatively smaller genome size (i.e., 280 Mbp), having eight autosomes in addition to one sex chromosome. Its other characteristics include its ability to grow quickly (it is frequently an invasive weed used in greenhouses) as well as its ability to propagate vegetatively through gemma cups (Poveda, 2020).

8.2.2 THE GYMNOSPERMS

Gymnosperms are primarily known for their dioecy, which is found in 64.6% (667) of all known species. Among gymnosperms, four families possess varying percentages of dioecy (i.e., 29.6% of Cupressaceae), 94.9% (Podocarpaceae), 5.4% (Araucariaceae), and 93.7% (Taxaceae), while six families are considered exclusively dioecious, namely, Gnetaceae, Cycadaceae, Ephedraceae, Zamiaceae, Welwitschiaceae, and Ginkgoaceae. Moreover, dioecy has disappeared in the *Pinaceae* (with the exception of *Pinus johannis*) and *Sciadopityaceae* families (Ohri and Rastogi, 2020). The number of dioecious species that are most likely possessing sex chromosomes is significantly smaller than the number of species that have sex chromosomes that can be identified cytologically. Consequently, more than half of these cases have homomorphic sex chromosomes since they haven't had enough time to differentiate morphologically. Furthermore, it has been suggested that gymnosperms, and cycad gametophytes in particular, have a low probability of producing heteromorphic sex chromosomes due to their larger size, longevity, and degree of complexity, which protects them from Muller's ratchet. Heteromorphism has mostly been found in relation to the satellite chromosomes, with the exception of the notable variation in sex chromosome size observed in *Cycas revoluta* (Rastogi and Ohri, 2020). Fluorescence in situ hybridization is therefore necessary for additional confirmation of these cases. In most gymnosperms, sex differentiation is still in its early stages and is caused by variations in cytosine methylation levels between the sexes because epigenetics play a key role in regulating sex differentiation. Since heterochromatic proteins bind to cytosine methylation sites, they can be distinguished between the sexes using G-banding under a light microscope. It is clear from this that there is still much to learn about the mechanism of sex differentiation in gymnosperms, and that a wide range of phylogenetically distinct species may be used to broaden the field. This suggests that dioecy has evolved more than once, giving rise to numerous additional sex-determining systems that still need to be identified in order to determine their precise mechanism of origin. As a result, additional research is required in various fields, including marker-related sex identification, chromosome banding, in situ fluorescent hybridization, functional genomics, cytosine methylation, and flow cytometry, and so forth. Hence, the challenge lies in techniques that distinguish between the sexes in a wide variety of species when cytology is not an option. Therefore, the development of sex-specific DNA markers offers a solution to the problem of identifying sexes during the vegetative stage, which is crucial for nursery trade conservation methods (Yang et al., 2022).

8.2.3 THE ANGIOSPERMS

Around 120,000 species of flowering plants are known to exist; of these, 72% produce bisexual flowers and 28% produce unisexual flowers. Roughly 10% of flowering species of plants are monoecious or absolutely dioecious (7% and 4%, respectively). While 10% of species have both bisexual and unisexual flowers, 7% of species exhibit intermediate types of sexual dimorphism, such as androdioecy and gynodioecy. Female or male flowers are produced on distinct individual plants in dioecious species, while female and male flowers are produced on a single plant but on different branches in monoecious species. Despite the fact that fewer species produce unisexual flowers than bisexual ones, some widely recognized and strategically significant plants are either monoecious or dioecious (Scutt, 2021). For instance, the monoecious species include *Cucumis melo* (melon), *Zea mays* (maize), *Ficus carica* (fig), and *Cucumis sativus* (cucumber), while *Spinacia oleracea* (spinach), *Humulus lupulus* (hop), *Asparagus ocyinalis* (asparagus), *Mercurialis annua* (mercury), and *Cannabis sativa* (hemp) are the dioecious species. Among the many mechanisms that have evolved to support outcrossing in angiosperms are the maturation of male or female organs at different times, structural avoidance, sporophytic or gametophytic self-incompatibility, monoecy, and eventually dioecy. The first step regarding the evolution of plant sex chromosomes is the appearance of female or male sterile mutations that lead to the formation of unisexual flowers. Numerous male sterile mutants found in domesticated agricultural crops, along with the presence of unisexual species within 75% of families of angiosperms, indicate that these mutations occur frequently as well as repeatedly in a variety of plant species (Malabadi et al., 2023). Moreover, a large number of genes play different roles in the development of carpels and stamens, and mutations in any of the many regulatory genes can cause the deletion or loss of function of either male or female organs. In dioecious flowers, female sterile mutation may arise from the initial phases of the carpel's inception and remain at the final phase of megaspore formation; likewise, male sterility may arise from the initial phase of the stamen primordia's first appearance and remain at the final phase of microspore formation (Cronk, 2022).

8.3 MODES OF SEX DETERMINATION

In plants, sex determination is categorized into the three major types as described below (Muyle et al., 2022):

1. Genetic sex determination (GSD) with male heterogamety that is, male (XO or XY); female (XX) and female heterogamety that is, male (ZZ); female (ZW).

2. Environmental sex determination (ESD).
3. Hormonal sex determination (HSD).

8.3.1 GENETIC SEX DETERMINATION (GSD)

Genetic sex determination (GSD) is a method that determines a person's sex based on the presence or absence of one or more important genetic markers. Plants' sex chromosomes and sex inheritance are very comparable to those of mammals (Mank, 2022). Males and females of separate-sex plants frequently produce a few blooms of the opposing sex, exhibiting "leaky" or "inconstant" sex expression (Cossard and Pannell, 2021). The degree and frequency of such inconsistency may indicate lingering hermaphrodite sex allocation following an evolutionary shift from single to mixed sexes (Spigler and Ashman, 2012). Additionally, a potential precursor to the end of dioecy and return to hermaphroditism is sex inconsistency. Monoecy and androdioecy have developed from dioecy in polyploid populations of the *Mercurialis annua* (Euphorbiaceae) species complex (Obbard et al., 2006).

Genotypes in the loci directly responsible for the formation of male and/or female features are referred to as genetically regulated sex determination. In the homozygous condition, one allele at each of these sites is useless. An allelic difference in only one locus results in gynodioecy (a difference only in the male determining locus) and also in androdioecy (a difference only in the female determining locus). Gynodioecy (a difference only in the male determining locus) results from an allelic variation in just one locus, whereas a difference in both loci would generate a fully dioecious state. Thus, sex chromosome evolution is thought to require the two loci that control sex determination (Charlesworth, 2013).

Most plants under study have heterozygous males, or male heterogamety (XX females, XY males). At that time, the chromosomes are clearly distinct (maybe half of the plants have distinct sexes) (Ming et al., 2013). There are several types of evidence that point to the participation of two loci in determining sex. Crosses between dioecious plants and allied hermaphrodite or monoecious species provide some of the data (Pannell and Gerchen, 2018). Cytological investigations have provided direct evidence of Y chromosomal deletions in *Silene dioica* and *S. latifolia.*

Both plants have three functionally distinct Y chromosomal sections: the suppressor of feminization (SuF) area, two regions containing factors regulating late and early development, and one region containing the SuF region (Fruchard, 2018). In these kinds of organisms, the majority of the Y chromosome does not undergo recombination during male meiosis, and X and Y pairing occurs only at the tips (Kauppi et al., 2018).

The sex determination process describes the emergence of an incipient sex chromosomal system, or a section of the chromosome that seldom recombines and contains the genes that determine sex (Beukeboom and Perrin, 2014). A recessive male sterility allele is carried by the female haplotype, whereas female sterility alleles would be carried by the dominant male-determining chromosome (Úbeda et al., 2015). Y chromosome degeneration and sex chromosomal evolution are closely related. The majority of what is now known about the development of Y chromosome characteristics originates from theoretical research on the evolution of genomic regions with low recombination. In the event that the rate of harmful mutations in such places is high enough, they are susceptible to several mechanisms (Abbott et al., 2015). To demonstrate various evolutionary stages, Westergaard (Charlesworth, 2018) divided plant chromosomes related to sex into three categories. The first stage possesses a viable YY genotype, which denotes the earliest stage of sex chromosome evolution, in which the Y solely differs from the X in its sex determination genes. The symbols for this syndrome are seen in asparagus and *Ecballium*. The Y chromosome is crucial in determining sex; white campion and papaya are symbols for this condition. The second stage has an inviable YY genotype. In the third stage, sex is defined by the X: autosome ratio, as demonstrated by sorrel, and the Y chromosome is unimportant (Ming et al., 2007). The Y chromosome is predicted to undergo degeneration upon the emergence of a sex-determining gene due to its reduced population size (where each male and female has three copies of the X chromosome and only one copy of the Y chromosome) and reduced recombination, which occasionally occurs between the Y and X chromosomes; thus, these factors reduce the effective population size of Y chromosomes and hence reduce the efficacy of selection, which can result in its degeneration (Meisel, 2020). Recombination reducing selection may result from sexual antagonism (SA), in which distinct SA alleles confer greater fitness in females and males. For instance, decreased recombination between the SA locus and the sex-determining region may link a male-benefit (female-detriment) allele to the Y chromosome, limiting the allele's advantages to males and shielding females from its disadvantages (Furman et al., 2020). Recombination frequently stops due to inversions that isolate progressively larger portions of the Y chromosome (Charlesworth et al., 2005). Plant sex chromosomes have been shown to be driven by two autosomes (Bergero and Charlesworth, 2009; Filatov, 2005; Nicolas et al., 2005). Theoretically, sex chromosomes emerge when at least two closely related sex-determining genes coexist; these genes are known as the "two-gene model," which includes a female sterility mutation (dominant in X/Y systems, recessive in Z/W systems) and a male sterility mutation (recessive in X/Y systems, dominant in Z/W systems) (Vyskot and Hobza, 2015).

The majority of our knowledge regarding the evolution of sex chromosomes is derived from a small number of ancient animal model organisms; that is, *Arabidopsis thaliana*. But unlike animals, the genes responsible for determining a plant's sex differ greatly from those responsible for an animal's (Charlesworth, 2019). There are plant species that exhibit different degrees of X-Y differentiation, which may correspond to distinct phases of the sex chromosomes' evolution (Furman et al., 2020). The distinctive feature of sex chromosomes is that each member of the pair possesses genetic material that varies from the other by a small amount, providing a means of identifying an individual's sex. This is caused by three different systems (Ohno, 2013). (i) Female heterogametic system, which is seen in some willows, such as *Salix suchowensis*, where females have two different sex chromosomes (ZW) and males are homogametic (ZZ) (Hou et al., 2015); (ii) male heterogametic system, as in *Silene latifolia* (Bernasconi et al., 2009), in which females are homogametic (XX) and males have two separate sex chromosomes (XY); and (iii) independent haplo-diploid phase that is, in *Marchantia polymorpha* (Yamato et al., 2007), sex in the haploid gametophytes can be determined by a UV system (females U, males V), and the diploid sporophyte is then heterogametic (UV) in their life cycle (Yamato et al., 2007). The following diagram shows different sex chromosome systems in various plant species: (i) male heterogamety (XY system) (i.e., *Silene latifolia*); (ii) female heterogamety (ZW) (i.e., *Salix suchowensis*); and (iii) haplo-diploid system (UV) (i.e., *Marchantia polymorpha*), showing U as maternal and V as paternal sex chromosomes (Muyle et al., 2017) (Figure 8.1; Table 8.2).

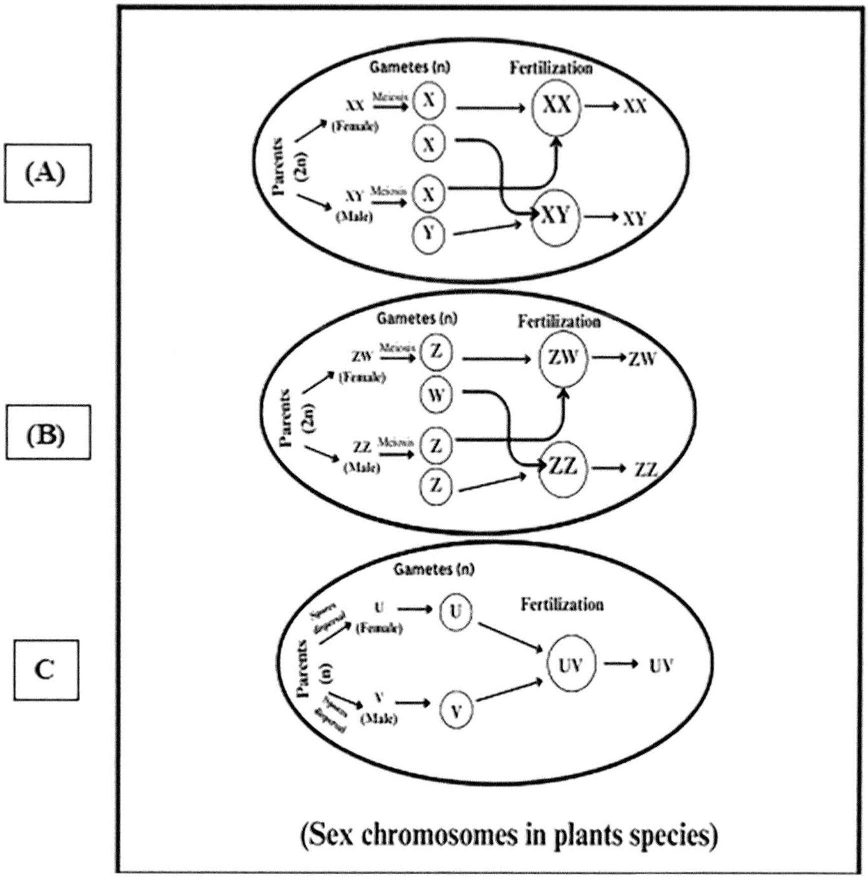

FIGURE 8.1 Sex chromosomes in plant species.

8.3.2 ENVIRONMENTAL SEX DETERMINATION

Environmental sex determination (ESD) is an approach by which sex can be determined after conception based on the environment, instead of being genetically fixed (Liu et al., 2021). Thus, it is expected that selection favors ESD when three criteria are fulfilled: (i) the habitat is patchy, with females performing more effectively compared to males in some patches and *vice versa* (Juvany et al., 2015); (ii) individuals have very little influence regarding the patch they settle on; and (iii) the environment significantly impacts life-long fitness. Various factors contribute to patchiness in the environment, such as competition for local mates, availability of resources, as well as rates of

TABLE 8.2 Sex Chromosomes UV, XY, ZW of Different Plant Species and Their Role in Sex Determination

Phylum	Class	Family	Scientific Name	Sexual System		Mechanism of Sex Determination	References
				Male (M)	Female (F)		
			Bryophytes				
Hepaticophyta (Liverworts)	Marchantiopsida	Marchantiaeceae	Umbrella liverwort (*Marchantia polymorpha*)	V	U	In haploid gametophyte, V (male) while U (female)	Goswami (2020)
			Marchantia inflexa	V	U	In haploid gametophyte, V (male) while U (female)	Marks et al. (2019)
		Sphaerocarpaceae	Texas balloonwort (*S. Texanus*)	V	U	In haploid gametophyte, V (male) while U (female)	Singh et al. (2023)
			S. donnellii	V	U	Same as above	Jibran (2022)
	Jungermanniopsida	Frullaniaceae	Dilated Scalewort (*Frullania dilatata*)	V	U	In haploid gametophyte, V (male) while U (female)	Sousa et al. (2021)
Bryophyta (Mosses)	Bryopsida	Ditrichaceae	fire moss (*Ceratodon purpureus*)	V	U	Same as above	Carey and Harkess (2021)
			Gymnosperms				
Cycadophyta	Cycadopsida	Cycadaceae	Sago Palm (*Cycas revoluta*)	XY	XX	X to autosome ratio	Rastogi and Ohri (2020)
			Siamese Cycad (*Cycas siamensis*)	XY	XX	Same as above	Ohri and Rastogi (2020)
		Zamiaceae	Natal grass cycad (*Stangeria eriopus*)	XY	XX	Same as above	Ohri and Rastogi (2020)
Gnetophyta	Gnetopsida	Ephedraceae	Shrubby horsetail (*Ephedra foliate*)	XY	XX	Same as above	Ohri and Rastogi (2020)

TABLE 8.2 *(Continued)*

Phylum	Class	Family	Scientific Name	Sexual System		Mechanism of Sex Determination	References
				Male (M)	Female (F)		
Angiosperms							
Magnoliophyta	*Caryophyllopsida.*	Caryophyllaceae	*Silene dioica*	XY	XX	Active Y system	Balounova et al. (2019)
			S. latifolia	XY	XX	–	Martin et al. (2019)
			S. diclinis	XY	XX	–	Kejnovsky et al. (2010)
			*S. pseudotite*s	XY	XX	–	Howell et al. (2009)
			S. colpophylla	XY	XX	–	Muyle et al. (2012)
			S. oleracea	XY	XY	Active Y system	Martin et al. (2019)
	Magnoliopsida	Polygonaceae	*Rumex angiocarpus*	XY	XX	–	Garcia et al. (2023)
			R. suffruticosus	XY	XX	–	Razumova et al., 2023
			R. graminifolius	XY	XX	–	Garcia et al., 2023
			R. acetosella	XY	XX	X to autosome ratio	Jesionek et al. (2021)
			R. papillaris	XY	XX	Active Y system	Garcia et al. (2023)
			R. hastatulus	XY	XX	X to autosome ratio	Joachimiak et al. (2022)
	Rosidsae	Rosaceae	*Fragaria chiloensis*	ZZ	ZW	Active W system	Tennessen et al. (2016)
			F. virginiana	ZZ	ZW	–	Goldberg et al. (2010)
			F. moschata	ZZ	ZW	–	Tennessen et al. (2016)
	Arecidae	Arecaceae	*Phoenix reclinata*	XY	XX	X to autosome ratio	Torres et al. (2018)
			P. atlantica	XY	XX	–	Cherif et al. (2016)
			P. dactylifera	XY	XX	–	Ali et al. (2017)

mortality specific to each sex. ESD is taxonomically dispersed throughout animals (in which sequential hermaphroditism is the most common form) and plants (frequently through sexual lability) (Tannenbaum et al., 2019). Although genetic sex determination (GSD) is considerably more common than ESD, ESD provides valuable insights for investigating evolutionary patterns of sex allocation in adaptable environments (Pawełkowicz et al., 2019).

Environmental sex determination (ESD) is analogous to the several kinds of sexual selection in terms of opposing selective forces on both females and males due to the reproductive cost involved (Van Doorn, 2014). Sexual selection is ubiquitous in various species, particularly in tree life; as a result, it frequently results in sexual dimorphism, that is, differences in size and appearance between sexes within the same species. Therefore, ESD takes into account different costs associated with female and male sex reproduction; thus, selective pressures have selected for flexibility across evolutionary time to determine sex in order to enhance fitness in a diverse environment (Crowley and Labonne, 2021). This is particularly important for organisms that cannot relocate to a new environment, such as sessile organisms. For instance, in plants, the sexual function of females requires greater energy for reproduction. Once fertilized, they use a significant amount of stored energy to produce sporophytes, fruits, or seeds. On the other hand, males only have to produce sperm (or sperm-bearing structures such as pollen produced by seed plants or antheridia produced by seedless plants) (Chauhan et al., 2014).

The environment's influence on sex determination is one of the most significant ways that result in variations in the ratio of male to female progeny. If the environment encountered during development is variable and has a sex-dependent effect on fitness, then it is expected that ESD is advantageous (Palmer et al., 2019). This phenomenon is widely observed in the animal kingdom, with ESD resulting in biased sex ratios of progeny, where environmental factors like temperature or parental factors are frequently involved. Plant species, on the other hand, are more plastic in terms of gender expression than animals, and they typically exhibit sex-allocation plasticity influenced by the environment, as well as variability in sex inconstancy at the time of flowering (Liu et al., 2021). The sex determination in plants, especially in seed plants (such as *Spinacia oleracea*), is rarely affected by the environment, and it is also unclear whether or not environmental variables interact with GSD systems to affect offspring sex ratios (Varga and Soulsbury, 2020).

The majority of fern species are homosporous and do not have sex chromosomes, with certain notable exceptions including the *Salviniales* type. All fern spores are able to develop into female, male, or hermaphrodite gametophytes based on their surroundings, as they do not possess genetic information specifying each sex (McCaw et al., 2020). Moreover, it has also been observed that breeding systems and ESD are related in numerous fern species, such as *Ceratopteris richardii* (Conway and Stilio, 2020). There are two main types of breeding systems seen in fern gametophytes: outcrossing and inbreeding. Antheridiogen, a chemical pheromone that regulates the sex of neighboring developing gametophytes, is released by female gametophytes to encourage outcrossing (Hornych et al., 2021). Thus, the transformation of neighboring asexual gametophytes into males is promoted by the antheridiogen secreted by females. This is adaptive as it increases the likelihood of outcrossing because males fertilize females with sperm instead of the females becoming bisexual or hermaphrodites and fertilizing themselves. However, even in the absence of fertilization, the female gametophyte may still develop into a hermaphrodite and self-fertilize in growth-promoting environments, which would eventually lead to a decline in inbreeding (Müller et al., 2020). Furthermore, the homosporous gametophyte of ferns is regulated by the abiotic environment according to the size-advantage model, just like in crocodilians. Additionally, gametophytes are shorter when they transform into males in stressful environments (which can include nutrient or overcrowding stress). On the other hand, gametophytes are larger and mature into females under more ideal growing conditions (Mateo et al., 2020). The phenomenon of sequential hermaphroditism is shown by plants, especially in angiosperms, that can alternate sexes repeatedly during their lives in order to maximize fitness during each flowering season (Barrett and Hough, 2013). However, ESD and sequential hermaphroditism are not mutually exclusive. For instance, the *Orchidaceae* family epiphyte *Catasetum viridiflavum* displays sequential hermaphroditism, in which the more mature, larger individuals have female inflorescences and those individuals who are younger and smaller have male inflorescences. However, light intensity also has a significant impact on sex expression (Bell et al., 2021). Regardless of size, individuals in bright sunlight are more likely to be female, and individuals in low light are more likely to be male. Moreover, individuals generate more ethylene, a commonly found plant hormone, in higher light, which encourages the development of female flowers (Leivesley et al., 2022).

Like ferns, moss gametophytes can be male, female, asexual, or hermaphrodite. As with vascular plants, moss gametophytes are capable of being either dioecious or monoecious, as opposed to homosporous ferns. The majority of studied dioecious species display GSD by sex determination system, especially UV sex chromosomes (Cronk, 2022). Low pH levels, insufficient light, and inadequate nutrient availability all favor male development in some monoecious moss species, which include *Splachnum ampullaceum*, that exhibit ESD during early development (Glime and Bisang, 2014). When a common plant hormone (i.e., auxin) or gibberellins compounds resembling fern antheridiogen are present, both female and male organisms make greater investments in their sexual structures (archegonia and antheridia). The process of determining gender in moss through environmental factors differs significantly from the spatial segregation of sexes found in sexually static lichen species, where sex ratios are mediated by the environment observed in moss patches (Valenzuela, 2008). The competitive advantage of female mosses results in differential survival rates between the sexes, which drives the spatial segregation of mosses. As a result, populations dominated by females are kept alive through minimal sexual and asexual reproduction. In contrast, the ESD is the dynamic development of males or females under various environmental conditions (Beukeboom and Perrin, 2014).

8.3.3 HORMONAL REGULATION IN SEX DETERMINATION

In many plant species, the regulation of sex determination is dependent on plant hormones (Gerashchenkov and Rozhnova, 2013; Khryanin, 2002; Yamasaki et al., 2005). There have been reports of cytokinin treatment feminizing male flowers in a number of species, such as *Pinus*, *Vitis*, *Spinacia*, and cannabis (Galoch, 1978; Luo et al., 2020; Ming et al., 2020). According to a different study, exogenous cytokinin treatment changed male *Vitis vinifera* flowers into hermaphrodite flowers (Coito et al., 2019). Grape sex determination is thought to be influenced by an allelic variation in a gene in the cytokinin metabolic process (CMP) (Men et al., 2021). Shy Girl, a cytokinin response regulator, has been identified as a putative sex determinant gene in kiwifruit that inhibits the development of carpels (Akagi et al., 2018). Moreover, determining the impact of exogenous hormones on flower development is obviously important when researching the role of hormones with respect to sex determination. However, little is known about how treatment with exogenous hormones at various developmental stages affects sex determination.

The transition of plants from the vegetative phase to flowering is thought to be facilitated by sexual hormones (Sheng et al., 2022). The female sex hormone functions in female plants, the male sex hormone functions in male plants, while both types of sex hormones function in monoecious plants. As hormones from plants and animals have been identified, their function in a variety of biological processes, most notably the expression of sex, has been clarified. The identification of the mechanisms governing the interaction between the genetic apparatus and the hormonal system, as well as their role in the differentiation of plant sexes, is one of the significant challenges facing modern biology (Chailakhyan and Khrianin, 2012), as these mechanisms pose important fundamental and practical implications.

Plant development and growth mechanisms, as well as sex differentiation, are often closely related processes (Diggle et al., 2011). The supporting data, however, are incredibly contradictory. Slow growth results in a predominance of maleness, according to some authors, while rapid growth enhances femaleness in plants. In this case, the rapid progression of developmental stages in plants is advantageous for the expression of female sex (Weyers and Paterson, 2001). Some authors show evidence of general growth retardation-induced plant feminization. Since phytohormones are essential for controlling plant development, growth, and sexuality, it has been proposed that phytohormones' impact on plant growth rate acts as a mediating factor in their effect on sex expression (Gerashchenkov and Rozhnova, 2013). In fact, the elongation of the shoots was always associated with increased maleness induced by gibberellin, whereas growth inhibition was linked to increased femaleness induced by cytokinin. Gibberellic acid (GA) had a greater effect on sex expression in the early maize (cv. Voronezhskaya-76), whereas cytokinin had a greater effect in the late maize (cv. Odesskaya-10) (Chailakhyan and Khrianin, 2012). It is unlikely that phytohormones' impact on sex expression is primarily due to developmental processes. In addition, not all cases showed a strong relationship between the amount of natural growth regulators and the rate of growth processes (Legris et al., 2019). This Figure 8.2 illustrates how gibberellins define male sex while cytokinins define female sex in plant species (Li et al., 2021).

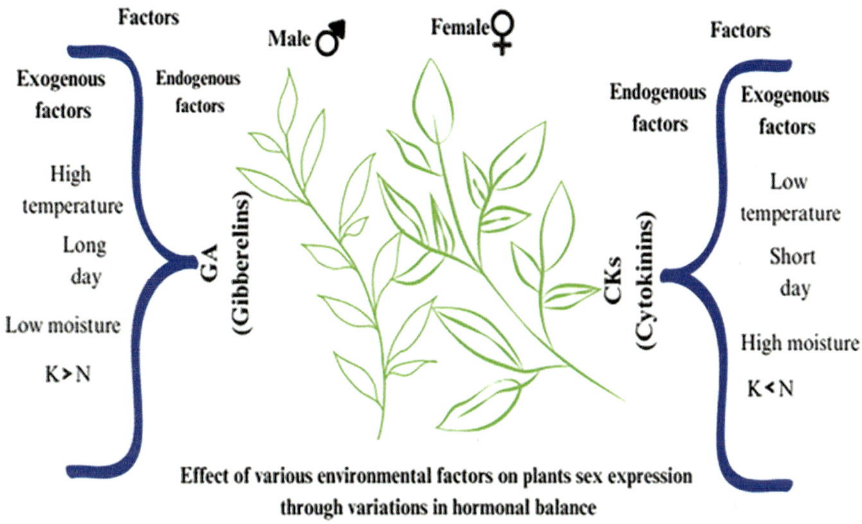

FIGURE 8.2 Environmental factors influencing sex expression through hormonal changes.

In contrast to flowering plants, especially ferns, more primitive plant forms also displayed the lability of sex expression. Thus, a spore in the fern *Ceratopteris richardii* has the ability to change either into a hermaphrodite with both archegonia and antheridia or into a male gametophyte with antheridia (Atallah et al., 2018). It was discovered that the hermaphrodite gametophytes secrete a sex hormone known as antheridiogen into the medium. This hormone's chemical structure is comparable to that of gibberellin molecules. When antheridiogen is present, spores germinate into male gametophytes, while those that do not acquire this hormone become hermaphrodite gametophytes. ABA can inhibit a particular effect of the antheridiogen. It has been noted that phytohormones play a part in the development, growth, and sex expression processes observed in the moss *Polytrichum commune* (Floriach-Clark et al., 2021). Various reports have shown that GA stimulated protonema growth, while abscisic acid (ABA) inhibited protonema growth but induced bud initiation of gametophore in young moss gametophytes on a nutrient medium that contained PGRs (plant growth regulators, i.e., gibberellins and cytokinins). Gametophores of two types were isolated: (i) those of the "gibberellin" type, which are elongated, that is, brown-colored internodes, small diameter of stem, shorter parenchymatous cells, and dentate well-developed egg-shaped leaves; and (ii) those of the "cytokinin" type, which are characterized by large parenchymal cells, poorly developed egg-shaped dentate leaves, a compact stem diameter, and green-colored internodes. Female individuals are primarily represented by the "cytokinin" type gametophore, while male individuals are represented by those of the "gibberellin" type gametophore. This suggests that, in comparison to many angiosperms, mosses and ferns are more versatile in terms of sex expression. However, the opinions of different authors regarding auxin's role in the regulation of plant sex differ significantly. Research revealed that treating plants with auxins changed the sex-to-sex relationship more gradually than treating them with cytokinins. Plants become significantly more feminine as a result of ethylene and its releaser, ethrel. This effect appears to be the consequence of the close relationship between ethylene and auxins and cytokinins. There is also conflicting and limited evidence regarding the impact of ABA on plant sex expression. It is important to acknowledge that ABA has the ability to induce femaleness in ferns as well as in both monoecious and dioecious plants with staminate and pistillate flowers. Gibberellins' masculinizing effect is eliminated by ABA; that's why its role in plant sex expression may be connected to the suppression of natural gibberellin activity. Furthermore, it is possible that the impact of ABA on sex expression is connected to the alteration in proline content, which is crucial for both the fertilization process and the transition of plants from the vegetative to the reproductive phase. This conclusion is supported by the observation that many plants' proline content increased significantly when ABA was applied exogenously through root application (Li et al., 2021).

Plant sex expression is known to be significantly influenced by ecological (environmental) factors. The environmental factors can alter the content of endogenous phytohormones, which regulate plant sex expression by influencing the activity of differentiating genes during the development of the generative organs. In plants, sex determination is largely dependent on the stage at which hormones begin to express. Cells and tissues cannot be prepared to receive the appropriate hormonal signals at an early stage (Gerashchenkov and Rozhnova, 2013). However, phytohormones are unable to alter sex expression once sex differentiation begins in the floral meristem. Moreover, plant sex determination is an essential developmental process that is especially significant from an economic perspective because the sexual phenotypes of crops that are important to the economy determine how those crops are bred and grown. While the majority of crop plants are not regarded as model systems, and the model angiosperm *Arabidopsis* cannot solve the issue of sex determination, the commercial benefits of modifying crop plant sexual phenotypes should continue to keep researchers focused on this field.

8.4 SUMMARY

The identification of a plant's sex is an essential developmental process that holds significant economic value, as the sexual phenotypes of crops with high commercial value determine the methods of breeding and cultivation. The economic benefits of modifying crop plants' sexual characteristics should keep researchers interested in this field, even though the majority of crop plants are not thought of as model systems and sex determination is not a problem that can be solved in the model angiosperm *Arabidopsis*. Numerous genetic and epigenetic elements, such as plant hormones and sex chromosomes, have contributed to the evolution of diverse sex-determining processes in angiosperms, including agricultural plants, as recent research on the subject has clearly illustrated. While the factors influencing sexual phenotype are heterogeneous, it will be necessary to clone the sex-determining genes from other plant species to ascertain whether the downstream master sex regulatory genes that dictate male or female development are shared or not.

Selecting to investigate sex determination in plants that belong to other major land plant lineages will enable the investigation of a number of more general developmental and evolutionary issues. How heterospory developed from homospory remains an open subject. One can test the theory that the transition from homospory to heterospory involved a heterochronic shift in the timing of expression of these genes from the gametophyte to the sporophyte generation by identifying the sex-determining genes in homosporous plants like *Ceratopteris* and analyzing the expression of potential homologous genes in closely related heterosporous species. How sex chromosomes developed in plants and whether comparable mechanisms resulted in different sex chromosomes in plants and animals are other unanswered concerns. Plants provide special mechanisms for studying sex determination in numerous ways. Numerous plant species have separately evolved the ability to produce unisexual blooms, suggesting that various novel mechanisms may be at work. Therefore, it seems unlikely that a single, overarching process accounts for plant sex determination. The investigation of the genetics, molecular biology, and biochemistry of the genes governing sexual determination in plants will lead to advances in our knowledge of sex determination. For these kinds of investigations, there are excellent model systems available for bisexual floral development (*Arabidopsis* and *Antirrhinum*), monoecy (maize), and dioecy (Silene, asparagus, and mercury). The key unanswered concerns are how sex determination genes work and whether or not they interact with homeotic genes that establish the sexual identity of primordia in floral organs. Although strong associations have been shown, the physiological relationship between hormone signaling and sexuality is not completely understood. Finally, new methods for controlling plant sexuality should emerge after the genes governing these processes are found, cloned, and investigated.

KEYWORDS

- **heterospory and homospory**
- **plant sexual development**
- **sex chromosomes**
- **sex determination**
- **sex differentiation**
- **sex-determining genes**
- **unisexual flower development**

REFERENCES

Abbott, J. K., Nordén, A. K., & Hansson, B. (2017). Sex chromosome evolution: Historical insights and future perspectives. *Proceedings of the Royal Society B: Biological Sciences, 284*(1854), 20162806.

Akagi, T., Henry, I. M., Ohtani, H., Morimoto, T., Beppu, K., Kataoka, I., & Tao, R. (2018). A Y-encoded suppressor of feminization arose via lineage-specific duplication of a cytokinin response regulator in kiwifruit. *The Plant Cell, 30*(4), 780–795.

Ali, H. B. M., Abubakari, A., Wiehle, M., & Krutovsky, K. V. (2017). Gene-specific sex-linked genetic markers in date palm (*Phoenix dactylifera* L.). *Genetic Resources and Crop Evolution, 65*, 1–10.

Atallah, N. M., Vitek, O., Gaiti, F., Tanurdzic, M., & Banks, J. A. (2018). Sex determination in *Ceratopteris richardii* is accompanied by transcriptome changes that drive epigenetic reprogramming of the young gametophyte. *G3: Genes, Genomes, Genetics, 8*(7), 2205–2214.

Ávila-Hernández, J. G., del Rosario Cárdenas-Aquino, M., Camas-Reyes, A., & Martínez-Antonio, A. (2023). Sex determination in papaya: Current status and perspectives. *Plant Science, 335,* 111814.

Balounová, V., Gogela, R., Cegan, R., Cangren, P., Zluvová, J., Šafář, J., Kováčová, V., et al. (2019). Evolution of sex determination and heterogamety changes in section *Otites* of the genus *Silene*. *Scientific Reports, 9*, 1045. https://doi.org/10.1038/s41598-018-37412-x.

Barrett, S. C., & Hough, J. (2013). Sexual dimorphism in flowering plants. *Journal of Experimental Botany, 64*(1), 67–82.

Bell, T. J., Bowles, M. L., Zettler, L. W., Pollack, C. A., & Ibberson, J. E. (2021). Environmental and management effects on demographic processes in the U.S. threatened *Platanthera leucophaea* (Nutt.) (*Orchidaceae*). *Plants, 10*(7), 1308.

Bergero, R., & Charlesworth, D. (2009). The evolution of restricted recombination in sex chromosomes. *Trends in Ecology & Evolution, 24*(2), 94–102.

Bernasconi, G., Antonovics, J., Biere, A., Charlesworth, D., Delph, L. F., Filatov, D., & Widmer, A. (2009). *Silene* as a model system in ecology and evolution. *Heredity, 103*(1), 5–14.

Beukeboom, L. W., & Perrin, N. (2014). *The Evolution of Sex Determination.* Oxford University Press.

Carey, S., Yu, Q., & Harkess, A. (2021). The diversity of plant sex chromosomes highlighted through advances in genome sequencing. *Genes, 12*(3), 381.

Chailakhyan, M. K., & Khrianin, V. N. (2012). *Sexuality in Plants and its Hormonal Regulation.* Springer Science & Business Media.

Charlesworth, D. (2013). Plant sex chromosome evolution. *Journal of Experimental Botany, 64*(2), 405–420.

Charlesworth, D. (2018). Mogens Westergaard's contributions to understanding sex chromosomes. *Genetics, 210*(4), 1143–1149.

Charlesworth, D. (2019). Young sex chromosomes in plants and animals. *New Phytologist, 224*(3), 1095–1107.

Charlesworth, D. (2022). The mysterious sex chromosomes of haploid plants. *Heredity, 129*(1), 17–21.

Charlesworth, D., Charlesworth, B., & Marais, G. (2005). Steps in the evolution of heteromorphic sex chromosomes. *Heredity, 95*(2), 118–128.

Chauhan, N. S., Sharma, V., Dixit, V. K., & Thakur, M. (2014). A review on plants used for improvement of sexual performance and virility. *BioMed Research International, 2014*, 868062.

Cherif, E., Zehdi-Azouzi, S., Crabos, A., Castillo, K., Chabrillange, N., Pintaud, J. C., Salhi-Hannachi, A., Glémin, S., & Aberlenc-Bertossi, F. (2016). Evolution of sex chromosomes prior to speciation in the dioecious *Phoenix* species. *Journal of Evolutionary Biology, 29*(8), 1513–1522. https://doi.org/10.1111/jeb.12887.

Chuck, G. (2010). Molecular mechanisms of sex determination in monoecious and dioecious plants. In *Advances in Botanical Research* (Vol. 54, pp. 53–83). Academic Press.

Coito, J. L., Silva, H. G., Ramos, M. J., Cunha, J., Eiras-Dias, J., Amâncio, S., & Rocheta, M. (2019). *Vitis* flower types: From the wild to crop plants. *PeerJ, 7*, e7879.

Connallon, T., Beasley, I. J., McDonough, Y., & Ruzicka, F. (2022). How much does the unguarded X contribute to sex differences in life span? *Evolution Letters, 6*(4), 319–329.

Conway, S. J., & Di Stilio, V. S. (2020). An ontogenetic framework for functional studies in the model fern *Ceratopteris richardii*. *Developmental Biology, 457*(1), 20–29.

Cossard, G. G., & Pannell, J. R. (2021). Enhanced leaky sex expression in response to pollen limitation in the dioecious plant *Mercurialis annua*. *Journal of Evolutionary Biology, 34*(2), 416–422.

Cronk, Q. (2022). Some sexual consequences of being a plant. *Philosophical Transactions of the Royal Society B, 377*(1850), 20210213.

Cronk, Q. (2022). The distribution of sexual function in the flowering plant: From monoecy to dioecy. *Philosophical Transactions of the Royal Society B, 377*(1850), 20210486.

Crowley, P. H., & Labonne, J. (2021). Theory of environmental sex determination: Trending populations in stressful environments. *Evolution, 75*(4), 794–805.

Diggle, P. K., Di Stilio, V. S., Gschwend, A. R., Golenberg, E. M., Moore, R. C., Russell, J. R., & Sinclair, J. P. (2011). Multiple developmental processes underlie sex differentiation in angiosperms. *Trends in Genetics, 27*(9), 368–376.

Filatov, D. A. (2005). Evolutionary history of *Silene latifolia* sex chromosomes revealed by genetic mapping of four genes. *Genetics, 170*(2), 975–979.

Floriach-Clark, J., Tang, H., & Willemsen, V. (2021). Mosses: Accessible systems for plant development studies. In *Model Organisms in Plant Genetics* (p. 17). IntechOpen.

Fruchard, C. (2018). *A Study of Sex Chromosomes and Sex Determination in Plants: Silene and Coccinia Systems Comparison* (Doctoral dissertation, Université de Lyon).

Fu, L., Wang, Q., Zhang, M., Zheng, Y., Wu, M., Lan, Z., & Lin, C. T. (2020). Electrochemical sex determination of dioecious plants using polydopamine-functionalized graphene sheets. *Frontiers in Chemistry, 8*, 92.

Furman, B. L., Metzger, D. C., Darolti, I., Wright, A. E., Sandkam, B. A., Almeida, P., & Mank, J. E. (2020). Sex chromosome evolution: So many exceptions to the rules. *Genome Biology and Evolution, 12*(6), 750–763.

Galoch, E. (1978). The hormonal control of sex differentiation in dioecious plants of hemp (*Cannabis sativa*): The influence of plant growth regulators on sex expression in male and female plants. *Acta Societatis Botanicorum Poloniae, 47*(1, 2), 153–162.

Garcia, S., Janousek, B., Pascual-Díaz, J. P., & Renner, S. S. (2023). Sex-chromosomes v. 2.0: A database of green plant species with sex chromosomes. *Chromosoma, 132*(1), 55–58.

Gerashchenkov, G. A., & Rozhnova, N. A. (2013). The involvement of phytohormones in plant sex regulation. *Russian Journal of Plant Physiology, 60*, 597–610.

Glime, J. M., & Bisang, I. (2014). Sexuality: Sex ratio and sex expression (*Bryophyte Ecology*, Vol. 1, Chap. 3–2). Michigan Technological University. https://digitalcommons.mtu.edu/bryo-ecol-subchapters/11 (accessed on 18 March 2025).

Goldberg, M. T., Spigler, R. B., & Ashman, T. L. (2010). Comparative genetic mapping points to different sex chromosomes in sibling species of wild strawberry (*Fragaria*). *Genetics, 186*(4), 1425–1433.

Goswami, H. K. (2020). Evolutionary lineages, inherently conserved within the genome of *Isoetes L.* (*Isoetaceae: Pteridophyta*) express in heterosporangia of *Isoetes pantii* Goswami & Arya: Brief review: A five decades' study. *Journal of Biotechnology & Bioinformatics Research, 2*(4), 1–8.

Hornych, O., Testo, W. L., Sessa, E. B., Watkins, J. E., Jr., Campany, C. E., Pittermann, J., & Ekrt, L. (2021). Insights into the evolutionary history and widespread occurrence of antheridiogen systems in ferns. *New Phytologist, 229*(1), 607–619.

Hou, J., Ye, N., Zhang, D., Chen, Y., Fang, L., Dai, X., & Yin, T. (2015). Different autosomes evolved into sex chromosomes in the sister genera of *Salix* and *Populus*. *Scientific Reports, 5*(1), 9076.

Howell, E. C., Armstrong, S. J., & Filatov, D. A. (2009). Evolution of neo-sex chromosomes in *Silene diclinis*. *Genetics, 182*(4), 1109–1115.

Intha, N., & Chaiprasart, P. (2018). Sex determination in date palm (*Phoenix dactylifera* L.) by PCR-based marker analysis. *Scientia Horticulturae, 236*, 251–255.

Jabbour, F., Espinosa, F., Dejonghe, Q., & Le Péchon, T. (2022). Development and evolution of unisexual flowers: A review. *Plants, 11*(2), 155.

Jesionek, W., Bodláková, M., Kubát, Z., Čegan, R., Vyskot, B., Vrána, J., & Hobza, R. (2021). Fundamentally different repetitive element composition of sex chromosomes in *Rumex acetosa*. *Annals of Botany, 127*(1), 33–47.

Jibran, R. (2022). The Renaissance and Enlightenment of Marchantia as a model system. *Plant Cell, 34*(10), 3512–3542.

Joachimiak, A. J., Libik-Konieczny, M., Wójtowicz, T., Sliwinska, E., & Grabowska-Joachimiak, A. (2022). Physiological aspects of sex differences and Haldane's rule in *Rumex hastatulus*. *Scientific Reports, 12*(1), 11145.

Juvany, M., & Munné-Bosch, S. (2015). Sex-related differences in stress tolerance in dioecious plants: A critical appraisal in a physiological context. *Journal of Experimental Botany, 66*(20), 6083–6092.

Kauppi, L., Jasin, M., & Keeney, S. (2012). The tricky path to recombining X and Y chromosomes in meiosis. *Annals of the New York Academy of Sciences, 1267*(1), 18–23.

Kejnovsky, E., & Vyskot, B. (2010). *Silene latifolia*: The classical model to study heteromorphic sex chromosomes. *Cytogenetic and Genome Research, 129*(1–3), 250–262.

Khryanin, V. N. (2002). Role of phytohormones in sex differentiation in plants. *Russian Journal of Plant Physiology, 49*, 545–551.

Kiełkowska, A., & Dziurka, M. (2021). Changes in polyamine pattern mediates sex differentiation and unisexual flower development in monoecious cucumber (*Cucumis sativus* L.). *Physiologia Plantarum, 171*(1), 48–65.

Lafuma, L., & Maurice, S. (2006). Reproductive characters in a gynodioecious species, *Silene italica* (*Caryophyllaceae*), with attention to the gynomonoecious phenotype. *Biological Journal of the Linnean Society, 87*(4), 583–591.

Legris, M., Ince, Y. Ç., & Fankhauser, C. (2019). Molecular mechanisms underlying phytochrome-controlled morphogenesis in plants. *Nature Communications, 10*(1), 5219.

Leite Montalvão, A. P., Kersten, B., Fladung, M., & Müller, N. A. (2021). The diversity and dynamics of sex determination in dioecious plants. *Frontiers in Plant Science, 11*, 580488.

Leivesley, J. A., Nancekivell, E. G., Brooks, R. J., Litzgus, J. D., & Rollinson, N. (2022). Long-term resilience of primary sex ratios in a species with temperature-dependent sex determination after decades of climate warming. *The American Naturalist, 200*(4), 532–543.

Leslie, A. B., & Bonacorsi, N. K. (2022). Understanding the appearance of heterospory and derived plant reproductive strategies in the Devonian. *Paleobiology, 48*(3), 496–512.

Li, H., Wang, L., Mai, Y., Han, W., Suo, Y., Diao, S., & Fu, J. (2021). Phytohormone and integrated mRNA and miRNA transcriptome analyses and differentiation of male between hermaphroditic floral buds of andromonoecious *Diospyros kaki* Thunb. *BMC Genomics, 22*, 1–19.

Liu, M., Korpelainen, H., & Li, C. (2021). Sexual differences and sex ratios of dioecious plants under stressful environments. *Journal of Plant Ecology, 14*(5), 920–933.

Liu, Y., Wang, S., Li, L., Yang, T., Dong, S., Wei, T., & Zhang, S. (2022). The *Cycas* genome and the early evolution of seed plants. *Nature Plants, 8*(4), 389–401.

Luo, Y., Pan, B. Z., Li, L., Yang, C. X., & Xu, Z. F. (2020). Developmental basis for flower sex determination and effects of cytokinin on sex determination in *Plukenetia volubilis* (Euphorbiaceae). *Plant Reproduction, 33*, 21–34.

Malabadi, R. B., Kolkar, K. P., Chalannavar, R. K., Munhoz, A. N. R., Abdi, G., & Baijnath, H. (2023). *Cannabis sativa*: Dioecious into monoecious plants influencing sex determination. *International Journal of Research and Innovations in Applied Science, 8*(7), 82–91.

Mank, J. E. (2022). Are plant and animal sex chromosomes really all that different? *Philosophical Transactions of the Royal Society B, 377*(1850), 20210218.

Marais, G. A., Gaillard, J. M., Vieira, C., Plotton, I., Sanlaville, D., Gueyffier, F., & Lemaitre, J. F. (2018). Sex gap in aging and longevity: Can sex chromosomes play a role? *Biology of Sex Differences, 9*, 1–14.

Marchant, D. B. (2019). Ferns with benefits. *American Fern Journal, 109*(3), 183–191.

Marchant, D. B., Chen, G., Cai, S., Chen, F., Schafran, P., Jenkins, J., & Chen, Z. H. (2022). Dynamic genome evolution in a model fern. *Nature Plants, 8*(9), 1038–1051.

Marchant, D. B., Sessa, E. B., Wolf, P. G., Heo, K., Barbazuk, W. B., Soltis, P. S., & Soltis, D. E. (2019). The C-Fern (*Ceratopteris richardii*) genome: Insights into plant genome evolution with the first partial homosporous fern genome assembly. *Scientific Reports, 9*(1), 18181.

Marks, R. A., Smith, J. J., Cronk, Q., Grassa, C. J., & McLetchie, D. N. (2019). Genome of the tropical plant *Marchantia inflexa*: Implications for sex chromosome evolution and dehydration tolerance. *Scientific Reports, 9*(1), 8722.

Martin, G., Carreel, F., Coriton, O., Hervouet, C., Cardi, C., Derouault, P., & D'Hont, A. (2017). Evolution of the banana genome (*Musa acuminata*) is impacted by large chromosomal translocations. *Molecular Biology and Evolution, 34*(9), 2140–2152.

Martin, H., Carpentier, F., Gallina, S., Gode, C., Schmitt, E., Muyle, A., & Marais, G. A. B. (2019). Evolution of young sex chromosomes in two dioecious sister plant species with distinct sex determination systems. *Genome Biology and Evolution, 11*(2), 350–361.

Mateo de Arias, M., Gao, L., Sherwood, D. A., Dwivedi, K. K., Price, B. J., Jamison, M., & Carman, J. G. (2020). Whether gametophytes are reduced or unreduced in angiosperms might be determined metabolically. *Genes, 11*(12), 1449.

McCaw, B. A., Stevenson, T. J., & Lancaster, L. T. (2020). Epigenetic responses to temperature and climate. *Integrative and Comparative Biology, 60*(6), 1469–1480.

Meisel, R. P. (2020). Evolution of sex determination and sex chromosomes: A novel alternative paradigm. *BioEssays, 42*(9), 1900212.

Meissner, S. T. (2021). Plant sexual reproduction: Perhaps the current plant two-sex model should be replaced with three- and four-sex models? *Plant Reproduction, 34*(3), 175–189.

Men, Y., Li, J. R., Shen, H. L., Yang, Y. M., Fan, S. T., Li, K., & Guo, X. W. (2021). *VaAPRT3* gene is associated with sex determination in *Vitis amurensis*. *Frontiers in Genetics, 12*, 727260.

Ming, R., Wang, J., Moore, P. H., & Paterson, A. H. (2007). Sex chromosomes in flowering plants. *American Journal of Botany, 94*(2), 141–150.

Ming, X., Tao, Y. B., Fu, Q., Tang, M., He, H., Chen, M. S., & Xu, Z. F. (2020). Flower-specific overproduction of cytokinins altered flower development and sex expression in the perennial woody plant *Jatropha curcas* L. *International Journal of Molecular Sciences, 21*(2), 640.

Mitiku, T., & Tujuba, C. (2022). Sex determination in plants and its contribution to genetic variability. *EAS Journal of Biotechnology and Genetics, 4*(4), 47–54.

Müller, N. A., Kersten, B., Leite Montalvão, A. P., Mähler, N., Bernhardsson, C., Bräutigam, K., & Fladung, M. (2020). A single gene underlies the dynamic evolution of poplar sex determination. *Nature Plants, 6*(6), 630–637.

Muyle, A., Marais, G. A., Bačovský, V., Hobza, R., & Lenormand, T. (2022). Dosage compensation evolution in plants: Theories, controversies and mechanisms. *Philosophical Transactions of the Royal Society B, 377*(1850), 20210222.

Muyle, A., Shearn, R., & Marais, G. A. (2017). The evolution of sex chromosomes and dosage compensation in plants. *Genome Biology and Evolution, 9*(3), 627–645.

Muyle, A., Zemp, N., Deschamps, C., Mousset, S., Widmer, A., & Marais, G. A. B. (2012). Rapid de novo evolution of X chromosome dosage compensation in *Silene latifolia*, a plant with young sex chromosomes. *PLOS Biology, 10*(4), e1001308.

Nicolas, M., Marais, G., Hykelova, V., Janousek, B., Laporte, V., Vyskot, B., & Monéger, F. (2005). A gradual process of recombination restriction in the evolutionary history of the sex chromosomes in dioecious plants. *PLOS Biology, 3*(1), e4.

Obbard, D. J., Harris, S. A., Buggs, R. J., & Pannell, J. R. (2006). Hybridization, polyploidy, and the evolution of sexual systems in *Mercurialis* (Euphorbiaceae). *Evolution, 60*(9), 1801–1815.

Ohno, S. (2013). *Sex chromosomes and Sex-Linked Genes* (Vol. *1*). Springer Science & Business Media.

Ohri, D., & Rastogi, S. (2020). Sex determination in gymnosperms. *The Nucleus, 63*, 75–80.

Palmer, D. H., Rogers, T. F., Dean, R., & Wright, A. E. (2019). How to identify sex chromosomes and their turnover? *Molecular Ecology, 28*(21), 4709–4724.

Pannell, J. R. (2017). Plant sex determination. *Current Biology, 27*(5), R191–R197.

Pannell, J. R., & Gerchen, J. (2018). Sex determination: Sterility genes out of sequence. *Current Biology, 28*(2), R80–R83.

Paterno, G. B., Silveira, C. L., Kollmann, J., Westoby, M., & Fonseca, C. R. (2020). The maleness of larger angiosperm flowers. *Proceedings of the National Academy of Sciences, 117*(20), 10921–10926.

Patiño, J., & Vanderpoorten, A. (2018). Bryophyte biogeography. *Critical Reviews in Plant Sciences, 37*(2, 3), 175–209.

Pawełkowicz, M. E., Skarzyńska, A., Pląder, W., & Przybecki, Z. (2019). Genetic and molecular bases of cucumber (*Cucumis sativus* L.) sex determination. *Molecular Breeding, 39*, 1–27.

Poveda, J. (2020). *Marchantia polymorpha* as a model plant in the evolutionary study of plant-microorganism interactions. *Current Plant Biology, 23*, 100152.

Rastogi, S., & Ohri, D. (2020). Chromosome numbers in gymnosperms—An update. *Silvae Genetica, 69*(1), 13–19.

Rastogi, S., & Ohri, D. (2020). Karyotype evolution in cycads. *The Nucleus, 63*, 131–141.

Razumova, O. V., Alexandrov, O. S., Bone, K. D., Karlov, G. I., & Divashuk, M. G. (2023). Sex chromosomes and sex determination in dioecious agricultural plants. *Agronomy, 13*(2), 540.

Renner, S. S., & Müller, N. A. (2022). Sex determination and sex chromosome evolution in land plants. *Philosophical Transactions of the Royal Society B, 377*(1850), 20210210.

Renner, S. S., Heinrichs, J., & Sousa, A. (2017). The sex chromosomes of bryophytes: Recent insights, open questions, and reinvestigations of *Frullania dilatata* and *Plagiochila asplenioides*. *Journal of Systematics and Evolution, 55*(4), 333–339.

Rodriguez-Granados, N. Y., Lemhemdi, A., Choucha, F. A., Latrasse, D., Benhamed, M., Boualem, A., & Bendahmane, A. (2017). Sex determination in *Cucumis*. In *Genetics and Genomics of Cucurbitaceae* (pp. 307–319). Springer.

Scutt, C. P. (2021). The origin of angiosperms. In *Evolutionary Developmental Biology: A Reference Guide* (pp. 663–682). Springer.

Sessa, E. B., & Der, J. P. (2016). Evolutionary genomics of ferns and lycophytes. *Advances in Botanical Research, 78*, 215–254.

Sheng, J., Li, X., & Zhang, D. (2022). Gibberellins, brassinolide, and ethylene signaling were involved in flower differentiation and development in *Nelumbo nucifera*. *Horticultural Plant Journal, 8*(2), 243–250.

Singh, S., Davies, K. M., Chagné, D., & Bowman, J. L. (2023). The fate of sex chromosomes during the evolution of monoicy from dioicy in liverworts. *Current Biology, 33*(17), 3597–3609.

Sousa, A., Schubert, V., & Renner, S. S. (2021). Centromere organization and UU/V sex chromosome behavior in a liverwort. *The Plant Journal, 106*(1), 133–141.

Spigler, R. B., & Ashman, T. L. (2012). Gynodioecy to dioecy: Are we there yet? *Annals of Botany, 109*(3), 531–543.

Sultanova, Z., Downing, P. A., & Carazo, P. (2023). Genetic sex determination, sex chromosome size, and sex-specific lifespans across tetrapods. *Journal of Evolutionary Biology, 36*(2), 480–494.

Suzuki, G. (2009). Recent progress in plant reproduction research: The story of the male gametophyte through to successful fertilization. *Plant and Cell Physiology, 50*(11), 1857–1864.

Tannenbaum, C., Ellis, R. P., Eyssel, F., Zou, J., & Schiebinger, L. (2019). Sex and gender analysis improves science and engineering. *Nature, 575*(7781), 137–146.

Tanurdzic, M., & Banks, J. A. (2004). Sex-determining mechanisms in land plants. *The Plant Cell, 16*, S61–S71.

Tennessen, J. A., Govindarajulu, R., Liston, A., & Ashman, T. L. (2016). Homomorphic ZW chromosomes in a wild strawberry show distinctive recombination heterogeneity but a small sex-determining region. *The New Phytologist, 211*(4), 1412–1423.

Torres, M. F., Mathew, L. S., Ahmed, I., Al-Azwani, I. K., Krueger, R., Rivera-Nuñez, D., Mohamoud, Y. A., Clark, A. G., Suhre, K., & Malek, J. A. (2018). Genus-wide sequencing supports a two-locus model for sex determination in *Phoenix*. *Nature Communications, 9*, 3969.

Úbeda, F., Patten, M. M., & Wild, G. (2015). On the origin of sex chromosomes from meiotic drive. *Proceedings of the Royal Society B: Biological Sciences, 282*(1798), 20141932.

Valenzuela, N. (2008). Sexual development and the evolution of sex determination. *Sexual Development, 2*(2), 64–72.

Van Doorn, G. S. (2014). Evolutionary transitions between sex-determining mechanisms: A review of theory. *Sexual Development, 8*(1–3), 7–19.

Varga, S., & Soulsbury, C. D. (2020). Environmental stressors affect sex ratios in sexually dimorphic plant sexual systems. *Plant Biology, 22*(5), 890–898.

Vyskot, B., & Hobza, R. (2015). The genomics of plant sex chromosomes. *Plant Science, 236*, 126–135.

Weyers, J. D., & Paterson, N. W. (2001). Plant hormones and the control of physiological processes. *New Phytologist, 152*(3), 375–407.

Wolf, D. E., & Takebayashi, N. (2004). Pollen limitation and the evolution of androdioecy from dioecy. *The American Naturalist, 163*(1), 122–137.

Wright, A. E., Dean, R., Zimmer, F., & Mank, J. E. (2016). How to make a sex chromosome? *Nature Communications, 7*, 12087.

Xirocostas, Z. A., Everingham, S. E., & Moles, A. T. (2020). The sex with the reduced sex chromosome dies earlier: A comparison across the tree of life. *Biology Letters, 16*(3), 20190867.

Yamasaki, S., Fujii, N., & Takahashi, H. (2005). Hormonal regulation of sex expression in plants. *Vitamins & Hormones, 72*, 79–110.

Yamato, K. T., Ishizaki, K., Fujisawa, M., Okada, S., Nakayama, S., Fujishita, M., & Ohyama, K. (2007). Gene organization of the liverwort Y chromosome reveals distinct sex chromosome evolution in a haploid system. *Proceedings of the National Academy of Sciences, 104*(15), 6472–6477.

Yang, Y., Ferguson, D. K., Liu, B., Mao, K. S., Gao, L. M., Zhang, S. Z., & Zhang, Z. X. (2022). Recent advances on phylogenomics of gymnosperms and a new classification. *Plant Diversity, 44*(4), 340–350.

Zhou, G., Yin, H., Chen, F., Wang, Y., Gao, Q., Yang, F., & Wan, Y. (2022). The genome of *Areca catechu* provides insights into sex determination of monoecious plants. *New Phytologist, 236*(6), 2327–2343.

CHAPTER 9

Chromosome and Mutations

MUHAMMAD ISHTIAQ,* MUBASHIR MAZHAR, MEHWISH MAQBOOL, MUHAMMAD AJAIB, TANVEER HUSSAIN, and MUHAMMAD WAQAS

Department of Botany, Mirpur University of Science and Technology (MUST), Mirpur, Azad Jammu and Kashmir (AJK), Pakistan

*Corresponding author

ABSTRACT

This chapter contains a comprehensive preview of mutations, causes of mutations, their types, and significance for natural flora and fauna, as well as cultivated species. This encompasses details about how mutations occur in nature or how they are induced in the desired species for targeted purposes. This chapter describes the various types of mutagens. The different types of mutagens are explicitly elaborated upon, along with their potential merits and demerits. Overall, it describes the significance of mutations, mutagens, and their role in nature and in man-made or cultivated plants and animals in various aspects of life.

9.1 MUTATION

A mutation is an alteration to an organism's deoxyribonucleic acid (DNA) sequence. Difficulties in DNA replication throughout cell division, exposure to mutagens, or viral infection can all cause mutations (Shen et al., 2022).

9.2 TYPES OF MUTATIONS BASED ON THE TYPES OF AFFECTED CELLS

The following categories of mutations can be used to categorize them based on the kinds of cells that the mutagen affects (Nanki et al., 2020).

Chromosomal and Genomic Structures and Functions, Volume 1: Chromosomes and Cell Division.
Tariq Ahmad Bhat and Jameel M. Al-Khayri (Eds)

DOI: 10.1201/9781779640253-9

9.2.1 SOMATIC MUTATIONS

These are genetic mutations that take place in the somatic (non-reproductive) cells within a living being. Due to the body's mechanisms for repair and compensation, a person may not be affected by a variety of somatic mutations. However, a somatic mutation that modifies a cell's pattern of CD can ultimately lead to the development of malignant cells or tissue (Vijg and Gossen, 1993).

9.2.2 GERM-LINE MUTATIONS

Gametes, reproductive cells that create gametes, or sex cells, can all experience these alterations. These mutations can be passed down to future generations and affect all of their cells (both somatic and germ-line). According to their mutagenicity in germ-line cells, many substances have been categorized and are included in the table (Campbell and Eichler, 2013).

9.2.3 TYPES OF MUTATIONS BASED ON THE TYPES OF EFFECT CAUSED

There are three known varieties of DNA mutations on the basis of the effect caused:

9.2.3.1 BASE MUTATIONS

Point mutation is the term for single-base substitution. These are the most frequent mutations that lead to silent, missense, or nonsense variations (Arnheim and Calabrese, 2009).

1. **Silent Mutations:** The replacement of a base at the codon's third position. As a result, there is a good chance that a codon that codes for the exact same sequence of amino acids will be produced. Therefore, there was no silent mutation brought on by an alteration in the gene sequence.

2. **Missense:** A base mutation that results in the production of a sequence of genes that codes for several distinct amino acids and ultimately results in the production of an alternative polypeptide sequence.

3. **Nonsense:** A nucleotide substitution that produces a sequence of genes that truncates translation, or a change that results in the production of a stop codon, ultimately leading to the formation of an ineffective protein, is referred to as a nonsense mutation.

9.2.3.2 DELETION

It is the elimination of any number of DNA base pairs, as the term implies. When any number of base pairs is lost or removed from the DNA, this might cause a frameshift (Strauss, 1999).

9.2.3.3 INSERTIONS

Adding more base pairs to the DNA might potentially cause frameshifts. Whether or whether the base pair additions occurred in multiples of three base pair will, however, determine whether or not there is a frameshift (Strauss, 1999).

9.3 MUTAGEN (CAUSATIVE AGENT OF MUTATION)

Mutagens are any substances that cause mutations. Mutagens can be either chemical, physical, or biological agents. The ability of a chemical to alter DNA base pairs or produce mutations is referred to as mutagenicity (Rodriguez and Loechler, 1993).

Deoxyribonucleic acid (DNA) makes up a living cell's genetic code. The polynucleotide known as DNA is made up of many different nucleotide units. Phosphate groups, sugars (particularly deoxyribose), and nitrogen-containing nucleobases (cytosine, guanine, adenine, or thymine) make up these units in turn. The usual arrangement of nucleic acid bases serves as the coding for all of a cell's genetic information. A mutagen alters the DNA's distinctive structure and nucleotide base order, which alters the proteins that are made from it. However, these alterations may be transmissible or not, depending on whether they take place in somatic or germline cells. Common mutagens include things like UV light, reactive oxygen species, X-rays, base analogs, alkylating agents, transposons, and so forth (Huang et al., 2011).

In layman's terms, mutagens are defined as substances that harm our genetic material, often DNA, and cause inherited or acquired genetic disorders. Agents that are mutagenic fulfill this role. According to the definition of mutagenic agents, the procedure entails developing mutagenic potential while also measuring and monitoring it. DNA is examined for spontaneously produced mutations to one or more probable mutagenic agents using transgenic non-human species (Goldman and Shields, 2003).

There are several substances that are known to cause mutations in the germ line, including those that have been proven via epidemiological research to result in germ-line heritable mutations. Chemicals that, according to mutagenicity experiments performed on mammalian somatic or germ cells, are believed to result in germ-line heritable mutations. Substances that have been shown in *in vivo* somatic cell genetic modification or somatic cell genotoxicity studies on animals to be capable of causing germ-line heritable mutations (Bara et al., 2021).

It is well known that mutagens enhance the frequency of mutations over the level of spontaneous mutation. However, it is crucial to realize that not every DNA damage or degradation qualifies as a mutation. The vast majority of the time, DNA polymerases remove the damaged or changed portion of DNA during the cell healing process. However, the mutations brought on by mutagens sometimes manage to evade this cell-repairing process, giving them the status of "permanent" DNA changes. The first mutagens to be identified were carcinogens, substances linked to cancer. In 1927, Hermann Joseph Muller conducted tests on fruit flies that led him to the discovery of the mutagenic capabilities of mutagens. He noticed that fruit flies can acquire genetic changes as a result of exposure to X-rays (Poetsch, 2020).

9.4 TYPES OF MUTAGENS

Physical, chemical, and biological mutagens are the three main categories of common mutagens found in nature (Auerbach, 1950).

9.4.1 PHYSICAL MUTAGENS

These include high-energy radiations like gamma, X, and alpha rays as well as temperature rise. Following are the types of physical mutagenic agents.

9.4.1.1 RADIATION OF HIGH ENERGY

Radiation with a high energy content is the main mutagenic agent identified in 1927. Mutagenic radiation includes X-rays, alpha rays, neutrons, and other ionizing and non-ionizing radiations. Typically, radiation causes direct damage to DNA or nucleotide structures, which may be fatal or only slightly fatal. One of the recognized mutagens that can result in damaging or sub-lethal alterations is the electromagnetic wave. In addition to DNA, a cell also contains proteins and lipids. Ionizing radiations like X-rays are primarily targeted at quickly proliferating cells.

9.4.1.2 TEMPERATURE

Another mutagen that causes DNA mutations in humans is temperature. The hydrogen bonds in the DNA are broken by the increase in temperature, causing it to denature into two single-stranded DNAs. In addition, DNA is harmed by high heat, which also destroys phosphodiester links.

9.4.2 CHEMICAL MUTAGENS

These include nitrous acid, alkylating compounds, acridines, and base analogues, among others. Following are the types of chemical mutagenic agents (Auerbach, 1967).

1. **Base Analogs:** The base analogs are elements that chemically or physically impersonate the DNA bases, such as pyrimidines as well as purines. When DNA is replicated, two frequent base analogs, bromouracil and aminopurine, are added into the structure as a replacement for systematic bases. Bromouracil has a Br group, which closely mimics thymine, in place of the methyl group of thymine (Khromov-Borisov, 1997).

2. **Alkylating Agents:** Common alkylating chemicals that add an alkyl radical to the DNA and harm it include ethyl nitrosourea,

poisonous gas, diethyl sulfate (DES), dimethyl nitrosamine (DMN), methylhydrazine, temozolomide, dacarbazine, busulfan, and ethyl ethane sulfate (EMS). These substances cause holes in the DNA strand and ethylation or methylation of nitrogen bases, which leads to base-pairing mistakes. Alkylated purine bases are removed by the process of depurination, which is not mutagenic. It could be fixed by the DNA repair mechanism (Grúz et al., 2022).

3. **Intercalating Agents:** One of the intercalating agents used in agarose gel electrophoresis is EtBr, or ethidium bromide. Other intercalating drugs, including daunorubicin, proflavine, and acridine orange, function by a similar mechanism to that of EtBr. When the molecules intercalate between the bases of DNA, the structure is disrupted. If it is incorporated during replication, it can cause a frameshift mutation. Furthermore, it will stop transcription (Ferguson and Denny, 2007).

4. **Metal Ions:** Because of how they interact with our DNA, metal ions are also harmful to it. Several of the common metal ions that result in mutations include iron, cobalt, cadmium, nickel, chromium, and cobalt. The reactive oxygen species (ROS) that the metal ions produce can damage DNA, obstruct DNA repair, lead to DNA hypermethylation, or the metal ions themselves can harm DNA directly.

5. **Other Chemical Mutagens:** Other mutagens that cause various mutations include ROS, benzene, gum elastic including rubber compounds, sodium azide, aromatic amines, alkaloids, deaminating chemicals, and polycyclic aromatic hydrocarbons (PAH).

9.4.3 BIOLOGICAL MUTAGENS

This category includes bacteria, transposons, and viruses. Following are the examples of biological mutagenic agents (Boone, 1991).

1. **Virus:** The common mutagens associated with infections, which can result in life-threatening health issues, are known to all of us. In our genome, viruses splice their DNA, interfering with how

DNA or genes typically function. Whenever the DNA has been introduced, viral protein is duplicated, translated, and again translated by the DNA as opposed to human protein. In a cell below, viral particles develop (Shapiro et al., 1984).

2. **Bacteria:** Some bacteria disrupt our DNA while causing inflammation in our internal organs. DNA damage and DNA fusion can be triggered by some bacteria (Touati, 2010).

3. **Transposons:** Transposons constitute less commonly recognized biological mutagens. Gene function is impacted by transposons, which are non-coding DNA sequences that migrate randomly throughout an immensely large genome.

9.5 MUTAGENIC EFFECT (MUTAGENESIS)

The mutagens are genotoxic, or destructive to our DNA, in a variety of ways. Some mutagens directly disrupt the DNA, while others do so indirectly. We are still unsure of the exact effects of any mutagen. At the chromosomal level, the mutagens can alter the number or structure of chromosomes. Chromosomal irregularities caused by mutagens include deletion, insertion, duplication, translocation, monosomy, and nondisjunction. The replication, transcription, and translation processes at the molecular level are likewise affected or dysregulated by the mutagens. At the molecular level, the mutagens cause various gene mutations that result in a protein having no function, a protein with altered function, or a protein with no function at all (DelRosso et al., 2023).

9.6 TESTING FOR MUTAGENICITY

Mutagenicity testing, which affects regulatory choices, aims to identify substances that may alter the genetic makeup of somatic as well as germ cells.

In contrast to the vast majority of other kinds of toxicity, genetic modifications sometimes take a long time to become apparent following exposure to the mutagen (da Silva Dantas et al., 2020; Sundar et al., 2018).

To adequately analyze the mutagenesis potential, three endpoints are often studied. As follows:

1. Gene mutation;
2. Structural chromosome aberrations;
3. Numerical chromosome aberrations.

The frequently used *in vitro* as well as *in vivo* studies are (EFSA Panel on Genetically Modified Organisms (GMO); Enslein, 1988; Mullins et al., 2021):

In vitro Testing	*In vivo* Testing
Ames test (bacterial reverse mutation test)	Rodent lethal dominant mutations tests
Chromosomal aberration test for mammals	Rats heritable translocation test
Gene mutation test for mammals	Specific locus test
Unscheduled DNA synthesis test	Exchange of sister chromatid test

9.7 ANTIMUTAGENS

It is a substance that has the power to inactivate mutagens, halt their actions, or otherwise stop them from interacting with DNA (Słoczyńska et al., 2014).

The enzymes involved in the DNA restoration, recombination, and replication processes may be directly or indirectly activated, inhibited, or inactivated.

These types of antimutagens include:

1. **Desmutagens:** These are substances that render mutagens partially or entirely inactive before the mutagen influences the genes via enzymatic or chemical interaction.

2. **Bio-Antimutagens:** These are real antimutagens that stop mutations after mutagens cause gene damage. They copy and repair the DNA that has been damaged by mutagens, which lowers the incidence of mutations.

9.8 TYPES OF ANTIMUTAGENIC AGENTS

The following are the main categories of antimutagenic substances (Bhattachar, 2011; Brockman et al., 1992; Słoczyńska et al., 2014):

9.8.1 VITAMINS

Doxorubicin-induced chromosomal abnormalities are shown to be resistant to the antimutagenic effects of vitamins C and E. When administered together with a pesticide, vitamin C (ascorbic acid) dramatically decreased the incidence of pesticide-induced mutations.

9.8.2 FLAVONOID

All flavones, including luteolin and kaempferol, along with additional phenolic hydroxyl group-containing flavonoids, show antimutagenic effects. It has been discovered that isoflavones, as well as flavonoid glycosides, possess strong antimutagenic effects. When investigated against 2-aminoanthracene, a substance that results in mutations caused by aflatoxin B1, hispidulin and hortensin have been shown to exhibit an antimutagenic effect. *Salmonella typhimurium* has been utilized to test the antimutagenic capacity of all these flavonoids against different mutagens. Citrus juice flavonoids contain anticarcinogenic and antimutagenic properties.

9.8.3 ANTHRAQUINONES

Anthraquinones from the *Aloe barbadensis* have been discovered to have antimutagenic effects. Chemicals that are chemically related to anthraquinones, such as anthrone, acridone, and xanthone, have antimutagenic properties, with anthrone being the most effective.

9.8.4 PHENOLIC COMPOUNDS

Some phenolic compounds, such as ellagic acid, which is found in foods like strawberries, raspberries, grapes, and walnuts, are said to be antimutagenic. Green tea's antimutagenic properties can prevent the frequency of sister chromatid exchanges brought on by cigarette smoking. Black tea's antimutagenic properties have been attributed to compounds like epicatechin, (-) epicatechin gallate, (-) epigallocatechins, along with (-) epigallocatechin gallate. Curcumin and eugenol, two phenolic compounds found in turmeric and clove, have been shown to block the mutagenicity of direct-acting mutagens such as N-methyl-N'-nitro-N-nitrosoguanidine. Eugenol also blocks tobacco-induced mutations in the Ames test.

9.8.5 CAROTENOIDS

It is well known that carotenoids can influence how promutagens activate. Canthaxanthin, carotene-8-apo-carotenal, and -carotene-8-apo-carotene methyl ester all show antimutagenic action, with a dose-dependent reduction in mutagenicity.

9.8.6 DITERPENOIDS

The diterpenoid-like chemical erythroxydiol, which was isolated from *Aquilaria agallocha*, had antimutagenic and anticancer properties.

9.8.7 COUMARINS

Umbelliferone is an example of antimutagenic substances that have been discovered in a variety of plant sources. Psoralen and osthol can lessen benzo[α]pyrene's mutagenicity.

9.8.8 TANNINS

The antioxidative qualities of certain tannins are essential for maintaining their anticarcinogenic as well as antimutagenic activities by avoiding cellular damage caused by oxidation, such as lipid peroxidation.

9.8.9 STEROIDS

The antioxidative qualities of certain tannins are essential for maintaining their anticarcinogenic as well as antimutagenic activities by avoiding cellular damage caused by oxidation, such as lipid peroxidation.

9.8.10 SAPONINS

Ginseng saponin metabolites injected into human gut microbes were found to be antigenotoxic towards benzo[α]pyrene-induced clastogenicity.

9.8.11 MARINE SUBSTANCES

Some secondary metabolites present in marine species have the potential to suppress the capacity of various mutagens to induce mutations in *Salmonella typhimurium*. Two antimutagenic compounds were found in a sea hare extract: elatol and obtusol.

9.8.12 MISCELLANEOUS SUBSTANCES

Ajoene and one of allicin's derivatives are two of the organosulfur compounds with significant antimutagenic action that are found in garlic extract. Several additional phytocompound subgroups, for example, piperine, have been identified. With regard to lead nitrate- and cyclophosphamide-induced micronuclei, including chromosomal abnormalities, lemongrass essential oil possesses antimutagenic activity. In central and southern India, this oil is used in the preparation of lemongrass tea.

9.8.13 PRODUCTS FROM FOOD AS ANTIMUTAGENS

Certain frequently consumed flavors, as well as vegetables, for example, turmeric, mustard, leafy green vegetables, and allium species, have been discovered to have antimutagenic and anticarcinogenic characteristics. Dietary desmutagens serve as tumor development inhibitors in the carcinogenesis process. Adding tomato, carrot, and spinach to the diet dramatically decreased the quantity of lymphocyte DNA damage, according to a human intervention experiment employing different carotenoids in vegetable products.

When tested on the *Salmonella Typhimurium* strains TA100 or TA1535, the garlic extract reduced the mutagenicity brought on by substances like sodium azide and N-methyl-N'-nitro-N-nitrosoguanidine, which function as direct mutagens. Casein showed a very high level of antimutagenic effect both *in vivo* and *ex vivo* in the DNA repair host-mediated test and liquid suspension experiment. Antimutagenic properties for yogurt with guava (*Psidium guajava*) have been identified. Researchers discovered that the antimutagenic actions of mushrooms were caused by a direct chemical interaction with the mutagens.

The microsomal activation-dependent genetic mutation of 2-acetamido fluorine is inhibited by extracts of asafetida and turmeric. Curcumin and eugenol, phenolic compounds found in turmeric as well as clove, and Indian cabbage leaf extract are said to be antimutagenic.

9.9 THE ANTIMUTAGENESIS MECHANISM

Anti-mutagenesis' primary processes can be generally summarized as follows (Shankel et al., 2013):

1. Chemical or enzymatic inactivation.
2. Suppression of the emergence of active species.
3. Scavenging.
4. Antioxidants.

9.9.1 CHEMICAL OR ENZYMATIC INACTIVATION

Numerous mutagens can be immediately neutralized by a variety of chemicals since they are highly reactive and affect proteins, enzymes, and DNA in addition to other cellular components.

Indole-3-carbinol and other cytochrome P-450 enzyme inducers and inhibitors have been connected to antimutagenic or anticarcinogenic properties.

For instance, benzyl isothiocyanate is an isothiocyanate. Antioxidants like 2, 3-tert-butyl-4-hydroxyanisole (BHA) can be mentioned, for example (Chan and Delucchi, 2000; Dixit and Kumar, 2018).

9.9.2 SUPPRESSION OF THE EMERGENCE OF ACTIVE SPECIES

Before they can form highly electrophilic species (the operational species) and interact with DNA, many genotoxic contaminants or carcinogens require metabolic or physiological activation. There is growing evidence that other organs, including the gastrointestinal tract (GIT), can also trigger metabolism, even though these activities often take place in the liver. As an illustration, the interaction of secondary and tertiary amines with nitrite in the stomach usually results in N-nitroso compounds.

9.9.3 SCAVENGING

To eliminate dietary mutagens, a number of desmutagens may adhere to or adsorb them. During this stage, the mutagen frequently remains unchanged but is unable to interact with DNA. Examples include several dietary fibers and chlorophyll.

9.9.4 ANTIOXIDANTS

Free radicals play a key role in the progression of cancer because they have the potential to damage DNA, cause mutagenicity, and induce cytotoxicity. ROS can impede DNA repair and create mutations, which inactivate some tumor suppressor genes and lead to cancer. Most free radicals can be promptly neutralized by a variety of antimutagenic drugs that have scavenger or free radical scavenging activities, particularly those with a brief half-life (Verma and Agarwal, 2004).

KEYWORDS

- **anthraquinones**
- **antimutagenic agents**
- **cell division**
- **deoxyribonucleic acid**
- **free radical scavenging**
- **reactive oxygen species**
- **tumor suppressor genes**

REFERENCES

Arnheim, N., & Calabrese, P. (2009). Understanding what determines the frequency and pattern of human germline mutations. *Nature Reviews Genetics, 10*(7), 478–488.

Auerbach, C. (1950). Differences between effects of chemical and physical mutagens. *Differences Between Effects of Chemical and Physical Mutagens, 22*(Suppl).

Auerbach, C. (1967). The chemical production of mutations: The effect of chemical mutagens on cells and their genetic material is discussed. *Science, 158*(3805), 1141–1147.

Bara, A. W., Braszewska, A., & Kwasniewska, J. (2021). DNA methylation—An epigenetic mark in mutagen-treated *Brachypodium distachyon* cells. *Plants, 10*(7), 1408.

Bhattachar, S. (2011). Natural antimutagens: A review. *Research Journal of Medicinal Plant, 5*(2), 116–126.

Boone, L. R. (1991). Biological mutagens. *Genetic Toxicology, 257.*

Brockman, H. E., Stack, H. F., & Waters, M. D. (1992). Antimutagenicity profiles of some natural substances. *Mutation Research/Fundamental and Molecular Mechanisms of Mutagenesis, 267*(2), 157–172.

Campbell, C. D., & Eichler, E. E. (2013). Properties and rates of germline mutations in humans. *Trends in Genetics, 29*(10), 575–584.

Chan, W. K., & Delucchi, A. B. (2000). Resveratrol, a red wine constituent, is a mechanism-based inactivator of cytochrome P450 3A4. *Life Sciences, 67*(25), 3103–3112.

da Silva Dantas, F. G., de Castilho, P. F., de Almeida-Apolonio, A. A., de Araujo, R. P., & de Oliveira, K. M. P. (2020). Mutagenic potential of medicinal plants evaluated by the Ames *Salmonella*/microsome assay: A systematic review. *Mutation Research/Reviews in Mutation Research, 786*, 108338.

DelRosso, N., Tycko, J., Suzuki, P., Andrews, C., Aradhana, Mukund, A., & Bintu, L. (2023). Large-scale mapping and mutagenesis of human transcriptional effector domains. *Nature, 616*(7956), 365–372.

Dixit, M., & Kumar, A. (2018). Mutagenesis, genetic disorders, and diseases. In *Mutagenicity: Assays and Applications* (pp. 1–34). Academic Press.

EFSA Panel on Genetically Modified Organisms (GMO), Mullins, E., Bresson, J. L., Dalmay, T., Dewhurst, I. C., Epstein, M. M., & Rostoks, N. (2021). *In vivo* and *in vitro* random mutagenesis techniques in plants. *EFSA Journal, 19*(11), e06611.

Enslein, K. (1988). An overview of structure-activity relationships as an alternative to testing in animals for carcinogenicity, mutagenicity, dermal and eye irritation, and acute oral toxicity. *Toxicology and Industrial Health, 4*(4), 479–498.

Ferguson, L. R., & Denny, W. A. (2007). Genotoxicity of non-covalent interactions: DNA intercalators. *Mutation Research/Fundamental and Molecular Mechanisms of Mutagenesis, 623*(1, 2), 14–23.

Goldman, R., & Shields, P. G. (2003). Food mutagens. *The Journal of Nutrition, 133*(3), 965S–973S.

Grúz, P., Yasui, M., Ukai, A., Horibata, K., Honma, M., & Sugiyama, K. I. (2022). Potent mutagenicity of an azide, 3-azido-1,2-propanediol, in human TK6 cells. *Mutation Research/Genetic Toxicology and Environmental Mutagenesis, 876*, 503475.

Huang, S. X., Jaurand, M. C., Kamp, D. W., Whysner, J., & Hei, T. K. (2011). Role of mutagenicity in asbestos fiber-induced carcinogenicity and other diseases. *Journal of Toxicology and Environmental Health, Part B, 14*(1–4), 179–245.

Khromov-Borisov, N. N. (1997). Naming the mutagenic nucleic acid base analogs: The Galatea syndrome. *Mutation Research/Fundamental and Molecular Mechanisms of Mutagenesis, 379*(1), 95–103.

Nanki, K., Fujii, M., Shimokawa, M., Matano, M., Nishikori, S., Date, S., & Sato, T. (2020). Somatic inflammatory gene mutations in human ulcerative colitis epithelium. *Nature, 577*(7789), 254–259.

Poetsch, A. R. (2020). The genomics of oxidative DNA damage, repair, and resulting mutagenesis. *Computational and Structural Biotechnology Journal, 18*, 207–219.

Rodriguez, H., & Loechler, E. L. (1993). Mutagenesis by the (+)-anti-diol epoxide of benzo[a]pyrene: What controls mutagenic specificity? *Biochemistry, 32*(7), 1759–1769.

Shankel, D. M., Hartman, P. E., Kada, T., Hollaender, A., Wilson, C. M., & Kuny, G. (2013). *Antimutagenesis and Anticarcinogenesis Mechanisms* (Vol. *39*). Springer Science & Business Media.

Shapiro, N. I., Marshak, M. I., & Varshaver, N. B. (1984). Mutagenic effects of DNA-containing oncogenic viruses and malignant transformation of mammalian cells. *Cancer Genetics and Cytogenetics, 13*(2), 167–179.

Shen, X., Song, S., Li, C., & Zhang, J. (2022). Synonymous mutations in representative yeast genes are mostly strongly non-neutral. *Nature, 606*(7915), 725–731.

Słoczyńska, K., Powroźnik, B., Pękala, E., & Waszkielewicz, A. M. (2014). Antimutagenic compounds and their possible mechanisms of action. *Journal of Applied Genetics, 55*, 273–285.

Strauss, B. S. (1999). Frameshift mutation, microsatellites, and mismatch repair. *Mutation Research/Reviews in Mutation Research, 437*(3), 195–203.

Sundar, R., Jain, M. R., & Valani, D. (2018). Mutagenicity testing: Regulatory guidelines and current needs. In *Mutagenicity: Assays and Applications* (pp. 191–228). Academic Press.

Touati, E. (2010). When bacteria become mutagenic and carcinogenic: Lessons from *H. pylori*. *Mutation Research/Genetic Toxicology and Environmental Mutagenesis, 703*(1), 66–70.

Verma, P. S., & Agarwal, V. K. (2004). *Cell Biology, Genetics, Molecular Biology, Evolution and Ecology: Evolution and Ecology*. S. Chand Publishing.

Vijg, J., & Gossen, J. A. (1993). Somatic mutations and cellular aging. *Comparative Biochemistry and Physiology Part B: Comparative Biochemistry, 104*(3), 429–437.

CHAPTER 10

Chromosome Movements During Cell Division

TANVEER HUSSAIN* and MUHAMMAD ISHTIAQ

Department of Botany, University of Azad Jammu and Kashmir, Bhimber, Azad Jammu and Kashmir (AJK), Pakistan

*Corresponding author

ABSTRACT

The movement of chromosomes toward the poles has been observed during mitosis by a new model. The key points of the new model are: (i) Kinetochore microtubules (KM) connected by filaments make up kinetochore spindle fibers (KSF). The KM is the site where microtubules (MTs) and filaments are interlinked to join at one point; (ii) In anaphase, motor molecules travel on the related filaments and toward the pole on the KM, which are embedded in a spindle matrix; (iii) In addition to the KSF pulling the kinetochore poleward, there are forces on the chromosome arms (CA) pushing them toward the pole in anaphase. These forces are independent and may originate from motor molecules in the spindle matrix; (iv) The KM adds subunits at the kinetochore and loses subunits from both the kinetochore and the pole. Motor molecules cause compression by pushing the KM toward the pole or into the kinetochore. During mitosis, duplicated DNA goes through a condensation or decondensation cycle. This is followed by the migration of sister chromatids to the metaphase plate. The nuclear envelope dissolves, allowing spindle assembly, chromosome segregation, and reformation of the nuclear envelope. Finally, genetically identical daughter cells are formed.

Chromosomal and Genomic Structures and Functions, Volume 1: Chromosomes and Cell Division.
Tariq Ahmad Bhat and Jameel M. Al-Khayri (Eds)

DOI: 10.1201/9781779640253-10

10.1 INTRODUCTION

The movement of chromosomes primarily appears during cell division (CD) after the active response of the spindle apparatus. There are three distinct kinds of microtubules (MT) in this spindle apparatus (SA): (i) the centriole-reflecting astral microtubules (AMT), which are present only in animal cells; (ii) continuous or inter-polar MTs that travel from one pole to the other; and (iii) kinetochore microtubules (KM), which start at the poles and end at the centromere.

Chromosomes biorient to opposing spindle poles during cell division, and chromatids partition equally between the two offspring cells, preventing aneuploidy—the gain or loss of chromosomes and its potentially harmful effects. Chromosome segregation and biorientation rely on interactions between MTs and kinetochores. Kinetochores cling to MTs, and chromosomes align with the metaphase plate during prometaphase. Early stages of chromosomal capture by spindle MTs can cause rapid poleward chromosome movements of up to 24 μm/min (Merdes and DeMey, 1990; Rieder and Alexander, 1990).

However, at the stage of prometaphase, chromosomes move toward the metaphase plate, and chromatids separate to move toward their respective poles during the anaphase stage. The interactions indicate a rapid transition to the mature shape of the kinetochore, where microtubules (MT) and their ends are embedded into them. They indicate interactions of basic means (Dong et al., 2007; McCleland et al., 2004; Rieder and Alexander, 1990). Currently, it has been observed that a KNL-1 protein is obtained as a result of protein assembly. Both the Ndc80 complex and the Mis12 complex support the connection between kinetochores and MTs (Cheeseman et al., 2006; Wei et al., 2007). The attachment of kinetochores to spindle fibers of MTs is directly mediated by the Ndc80 complex (Cheeseman et al., 2006; DeLuca et al., 2006; Wei et al., 2007).

It was also indicated previously that we showed the direct inhibition of the Ndc80 complex by antibody microinjection. At the anaphase stage, it led to the loss of most chromatid movement. However, before the prometaphase, chromosomes continuously bind to the mitotic spindle MTs in the same cells. They exhibited oscillatory movements, although common metaphase was impaired. Thus, these initial studies suggested that while Ndc80 complex-dependent mechanisms appear to be required for normal anaphase chromatid movement, other mechanisms are independent of the

Ndc80 complex. During the prometaphase stage, foster chromosomes were attached and moved (McClelland et al., 2009).

Pro-metaphase cell connections between kinetochores and MTs lead to fast poleward movements. Transient kinetochore-to-microtubule connections may be mediated by the neurotransmitter dynein (Rieder and Alexander, 1990). However, the majority of evidence for the involvement of dynein/dyactin in this pathway is indirect. Dynein accumulates on non-synonymous kinetochores and is depleted as initial lateral kinetochore interactions develop into end-on connections (King and Schroer, 2000). Both the speed and direction of chromosome translocation immediately after nuclear envelope degradation are consistent with the function of dynein and dyneactin. In the well-known model organism (*S. cerevisiae*), there are 6 kinesins and 6 dyneins. However, the capture and retrieval of long-range chromosomes is dependent on a minus-end directed Kinesin-14 (Kinesin-14) called Kar3p (Sproul et al., 2005). In *S. pombe*, Klp2p (also a member of the subfamily kinesin 14) aids in chromosome translocation toward the spindle pole body, apparently by depolymerizing kinetochore fiber MTs (Tanaka et al., 2007). It has recently been observed in budding yeast (as well as in vertebrates) that chromosomes can travel along lateral surfaces of the MTs.

A major obstacle in dissecting the role of dynactin function or dynein function in the process of mitosis is the transient nature, especially during the initial centromere attachment in the early events of chromosome capture and movement during prometaphase. Under normal circumstances, these initial interactions are quickly masked by the stronger Ndc80 complex-dependent binding of kinetochores to MTs.

To determine the potential role of dynactin/dynein in the movement of chromosomes during vertebrate cell mitosis, we first blocked the function of the Ndc80 complex and then studied the role of dynein/dynactin in a system where the direction and speed of chromosome movement can be clearly quantified. Our results show that rapid chromosome movement on MTs during early prometaphase is regulated by dynein/dynactin, but tethering mediated by the Ndc80 complex is required for metaphase association and chromosome movement during anaphase (Polley et al., 2023).

The mitotic spindle is a MT-based organelle responsible for ensuring accurate chromosome segregation during cell division (Mitchison and Salmon, 2001). This structure, which looks a lot like an American football, must be built by the cell before each round of division. Errors in its assembly can lead to uneven distribution of chromosomes in daughter cells. Known as aneuploidy, this condition is a hallmark of many cancers and a cause of birth defects if it occurs early enough in development. At the onset of mitosis in multicellular animal cells, disruption of the nuclear envelope releases signaling pathways previously limited by the intact interphase nucleus. One of the most important manifestations of this cell division process is spindle assembly, which is coordinated in part by signaling signals that form local gradients around chromatin. The chromatin-mediated Ran/import and CPC pathways are the best studied. Disruption of the nuclear envelope also allows nucleated MTs from the centrosome to enter the old nucleoplasmic space and "search" for attachment sites on chromosomes.

These sites, called kinetochores, provide functional controllers that couple the force generated by the spindle to the "captured" chromosomes. Additional spindle MTs are nucleated around chromosomes through the local release of spindle assembly factors (SAFs) from bound importins. The newly nucleated MTs are then integrated into the spindle by the coordinated action of motor proteins that can slide and orient the MTs relative to each other. Once the spindle is formed and all chromosomes are properly attached, the spindle assembly checkpoint (SAC) is satisfied, the cell enters anaphase, and the sister chromatids of the chromosomes' replication are separated by forces generated by the spindle (Ruben et al., 2023).

From many detailed drawings of the mitosis process, it was indicated by Walther Flemming that he deduced correctly but could not stabilize the chromosome sequence movement during the process of mitosis. Flemming divided mitosis into two main parts: a progressive phase, in which the chromosomes condense and arrange at the middle of the spindle fibers, and a regressive phase during which the sister chromatids detach. Our current understanding of mitosis has been uplifted by the latest advancements in light microscopy that allow researchers to monitor mitosis in living cells. The imaging of live cells reveals an extremely dynamic process that not only confirms Flemming's observations but can also be partially appreciated in the form of still images (Velten et al., 2013).

Today, mitosis consists of five stages. This division is based on the physical state of the chromosomes and their spindle fibers. The names of these stages are as follows: prophase, prometaphase, metaphase, anaphase, and telophase. Cytokinesis is the last physical cell stage of division, which appears after telophase and is therefore sometimes considered the sixth stage of mitosis. All stages of mitosis, as well as the adjacent stages of interphase and the preceding and following cell divisions, are important to understand.

10.1.1 PROPHASE

During this stage of mitosis, the recruited chromosomes condense and begin to undergo condensation that will continue until the metaphase stage is elaborated or appears. In most species, cohesin is largely removed from sister chromatid arms during prophase, allowing for the resolution of individual sister chromatids. However, cohesion is retained in the most restricted part of the chromosome, the centromere. During this stage, the spindle fibers also begin to form microtubules (MTs) as two pairs of centrosomes move away from the poles and start polymerization from the copied centrosomes (Firat-Karalar and Stearns, 2014).

10.1.2 PROMETAPHASE

This stage begins with the fragmentation of the nuclear envelope into many small vesicles, which will be divided between the coming daughter cells. An essential step in spindle assembly is the rupture of the nuclear membrane. Hence, the centrosome is seen outside the nucleus in animal cells; microtubules in the growing spindle cannot reach the chromosomes until the nuclear membrane ruptures (Potapova et al., 2011). Microtubules quickly combine and disassemble as they are produced from centrosomes, searching for binding sites at chromosome kinetochores, which are more complex sheet-like planar structures that assemble during pro-metaphase on the face of each sister chromatid at its centromere. As prometaphase proceeds, the chromosomes are pulled back and forth in opposite directions by MTs growing from the spindle poles until the polar forces are eventually balanced. The sister chromatids do not separate in this tug-of-war because they are held tightly together by adhesive residue left at their centromeres. At the end of prophase, chromosomes have bidirectional orientation, meaning that the kinetochores of sister chromatids are connected by MTs at opposite poles of the spindle (Potapova et al., 2011).

10.1.3 METAPHASE

The stage metaphase collects chromosomes at the equator of the spindle when the centromeres of all the cell's chromosomes are lined up. The metaphase stage is very helpful during cytogenetics because the chromosomes can be visualized easily at this stage. Moreover, at metaphase, with mitotic poisons such as colchicine, cells can be arrested experimentally very well. Microscopic video indicates that chromosomes temporarily stop movement during metaphase. The mechanism determines a complex checkpoint: either the spindle is properly assembled, but only cells assembled correctly with spindles can then enter the next stage of mitosis, named anaphase (Joglekar and Hunt, 2002).

10.1.4 ANAPHASE

The progression of the cell from metaphase to anaphase is marked by the sudden separation of sister chromatids. The main reason for the separation of chromosomes is the rapid degradation of cohesin molecules connecting sister chromatids by the protease separase. Two elaborative forms of movement appear during the anaphase stage of mitosis. As the first part of anaphase indicates, kinetochores (KMs) shorten and chromosomes move toward the spindle pole. During the second part of anaphase, the spindle poles separate as non-KMs pass each other. These latter movements are now thought to be catalyzed by motor proteins that connect microtubules (MTs) of opposite polarity and then "walk" toward the end of the MT (Asbury, 2017).

10.1.5 TELOPHASE AND CYTOKINESIS

The process of mitosis complete during last stage named as telophase, or the stage where chromosomes will appear at respective poles. This stage indicated that the nuclear membrane reformation and the decoding of chromosomes begin into their interphase structures. Telophase is followed by mitosis, or the breakdown of cytoplasm into two equal daughter cells (Choi et al., 2022). The daughter cells obtained from this process have identical genetic composition.

KEYWORDS

- **anaphase movement**
- **cell division**
- **chromosome segregation**
- **kinetochore microtubules**
- **mitosis**
- **motor proteins**
- **spindle matrix**

REFERENCES

Asbury, C. L. (2017). Anaphase A: Disassembling microtubules move chromosomes toward spindle poles. *Biology, 6*(1), 15.

Cheeseman, I. M., Chappie, J. S., Wilson-Kubalek, E. M., & Desai, A. (2006). The conserved KMN network constitutes the core microtubule-binding site of the kinetochore. *Cell, 127*(5), 983–997.

Choi, J., Clark, M. A., & Douglas, M. (2022). The cell cycle. *Biology 2e Part I* (2nd edn.).

DeLuca, J. G., Gall, W. E., Ciferri, C., Cimini, D., Musacchio, A., & Salmon, E. D. (2006). Kinetochore microtubule dynamics and attachment stability are regulated by Hec1. *Cell, 127*(5), 969–982.

Dong, J., Feldmann, G., Huang, J., Wu, S., Zhang, N., Comerford, S. A., Gayyed, M. F., Anders, R. A., Maitra, A., & Pan, D. (2007). Elucidation of a universal size-control mechanism in *Drosophila* and mammals. *Cell, 130*(6), 1120–1133.

Fırat-Karalar, E. N., & Stearns, T. (2014). The centriole duplication cycle. *Philosophical Transactions of the Royal Society B: Biological Sciences, 369*(1650), 20130460.

Joglekar, A. P., & Hunt, A. J. (2002). A simple, mechanistic model for directional instability during mitotic chromosome movements. *Biophysical Journal, 83*(1), 42–58.

King, S. J., & Schroer, T. A. (2000). Dynactin increases the processivity of the cytoplasmic dynein motor. *Nature Cell Biology, 2*(1), 20–24.

McCleland, M. L., Kallio, M. J., Barrett-Wilt, G. A., Kestner, C. A., Shabanowitz, J., Hunt, D. F., & Stukenberg, P. T. (2004). The vertebrate Ndc80 complex contains Spc24 and Spc25 homologs, which are required to establish and maintain kinetochore-microtubule attachment. *Current Biology, 14*(2), 131–137.

McClelland, S. E., Burrell, R. A., & Swanton, C. (2009). Chromosomal instability: A composite phenotype that influences sensitivity to chemotherapy. *Cell Cycle, 8*(20), 3262–3266.

Merdes, A., & De Mey, J. (1990). The mechanism of kinetochore-spindle attachment and poleward movement analyzed in PtK2 cells at the prophase-prometaphase transition. *European Journal of Cell Biology, 53*(2), 313–325.

Mitchison, T. J., & Salmon, E. D. (2001). Mitosis: A history of division. *Nature Cell Biology, 3*(1), E17–E21.

Polley, S., Müschenborn, H., Terbeck, M., De Antoni, A., Vetter, I. R., Dogterom, M., & Huis in't Veld, P. J. (2023). Stable kinetochore-microtubule attachment requires loop-dependent Ndc80-Ndc80 binding. *The EMBO Journal, 42*(13), e112504.

Potapova, T. A., Sivakumar, S., Flynn, J. N., Li, R., & Gorbsky, G. J. (2011). Mitotic progression becomes irreversible in prometaphase and collapses when Wee1 and Cdc25 are inhibited. *Molecular Biology of the Cell, 22*(8), 1191–1206.

Rieder, C. L., & Alexander, S. P. (1990). Kinetochores are transported poleward along a single astral microtubule during chromosome attachment to the spindle in newt lung cells. *The Journal of Cell Biology, 110*(1), 81–95.

Ruben, B. S., Brahmachari, S., Contessoto, V. G., Cheng, R. R., Junior, A. B. O., Di Pierro, M., & Onuchic, J. N. (2023). Structural reorganization and relaxation dynamics of axially stressed chromosomes. *Biophysical Journal, 122*(9), 1633–1645.

Sproul, L. R., Anderson, D. J., Mackey, A. T., Saunders, W. S., & Gilbert, S. P. (2005). Cik1 targets the minus-end kinesin depolymerase Kar3 to microtubule plus ends. *Current Biology, 15*(15), 1420–1427.

Tanaka, K., Kitamura, E., Kitamura, Y., & Tanaka, T. U. (2007). Molecular mechanisms of microtubule-dependent kinetochore transport toward spindle poles. *The Journal of Cell Biology, 178*(2), 269–281.

Velten, A., Wu, D., Jarabo, A., Masia, B., Barsi, C., Joshi, C., & Raskar, R. (2013). Femto-photography: Capturing and visualizing the propagation of light. *ACM Transactions on Graphics (ToG), 32*(4), 1–8.

Wei, R. R., Al-Bassam, J., & Harrison, S. C. (2007). The Ndc80/HEC1 complex is a contact point for kinetochore-microtubule attachment. *Nature Structural & Molecular Biology, 14*(1), 54–59.

Index

H

I

J

K

N

Q

R

S

T

U

V

For Product Safety Concerns and Information please contact our EU representative GPSR@taylorandfrancis.com Taylor & Francis Verlag GmbH, Kaufingerstraße 24, 80331 München, Germany